Diagnosis of
Mineral Disorders in Plants

Volume 2

Vegetables

by Alan Scaife and Mary Turner

new photographs by

Philip Wood
National Vegetable Research Station.
Wellesbourne, UK

General Editor: J. B. D. Robinson
Long Ashton Research Station, University of Bristol, UK

LONDON:
HER MAJESTY'S STATIONERY OFFICE

ISBN 0 11 240804 4

Contents

Introduction

Vegetable growing in Britain varies in scale from small back garden operations to enterprises covering thousands of hectares. At all levels of activity there is a desire to diagnose and rectify any disease or disorder which might reduce the yield or quality of the crop. Frequent changes in practice, increasing reliance on mineral sources of nutrients, and the regular introduction of new pesticides which may be marginally phytotoxic, mean that it is vital that the grower and advisor are able to recognise the symptoms of mineral imbalance, and to distinguish them from similar symptoms having other causes. We hope that this book will facilitate this process.

The principles and methods used in the diagnosis of mineral disorders have been dealt with in volume one, and the present volume therefore consists largely of photographic illustrations of disorders (mainly deficiencies), together with descriptive notes, on 16 temperate zone vegetable crops. The potato and tomato crop will be dealt with in succeeding volumes. Although the emphasis is necessarily on British crops and problems, we have made use of some pictures and references from elsewhere for the sake of completeness.

The reader confronted with an unhealthy crop will no doubt be tempted to turn directly to the photographs of that crop, and so the arrangement of the 'colour atlas' is such that each crop appears in alphabetical order of its common name. ('Brassicas' are grouped together, as are French and runner beans, which appear as 'Phaseolus'.) Within each crop, the order is: deficiencies of nitrogen (N), phosphorus (P), sulphur (S), potassium (K), calcium (Ca), magnesium (Mg), iron (Fe), manganese (Mn), zinc (Zn), copper (Cu), boron (B), molybdenum (Mo), followed by toxicities, ie effects of excess, of sulphur, manganese, boron, aluminium (Al) and chlorine (Cl). Deficiencies of the six major nutrients (N, P, S, K, Ca, Mg) are included for all crops except watercress, for which N, P and K only are shown: for micronutrients, the aim has been to include mainly those deficiencies known to have occurred in commercial practice somewhere in the world. As to toxicities, very few are shown: for many heavy metals, the toxicity symptoms are similar to those of iron deficiency: for others, the best source of information is probably Chapman (1966). The complete syndrome description for each disorder appears at the appropriate point, even for those few cases where no illustration is available. This is followed by a caption for each picture, including, if available, analytical values for the plants illustrated. Footnotes draw attention to 'lookalike' symptoms and provide references to more detailed information, such as accounts of field occurrences of the disorder in question.

Although it is natural to rely heavily on the pictures alone to provide a diagnosis, it is not often that this will prove unambiguous. The diagnostic process includes consideration of all available evidence, including the probability of observing a particular disorder in the circumstances prevailing, eg soil type, the susceptibility of the crop to that disorder and the possibility of the symptoms in question having arisen from other causes. These matters are dealt with, element by element, in chapter one, and much more thoroughly in volume one. This chapter also includes approximate 'critical concentrations' of each element in leaves (see below), and treatments, as recommended by the Agricultural Development and Advisory Service of the British Ministry of Agriculture, Fisheries and Food, for prevention and cure of each disorder.

Frequently neither the appearance of the crop nor the evidence about soil etc. is sufficient for a confident diagnosis, in which case some form of plant analysis will be necessary. This again is dealt with in volume one, but chapter two of this volume provides brief notes on how to collect samples for this purpose. It also explains the term 'critical concentrations' for interpreting the results of plant analysis, and gives details of a quick sap test which enables growers to monitor plant nitrate levels without the help of an analytical laboratory.

Many of the new photographs of major element and boron deficiencies resulted from two pot experiments done at the National Vegetable Research Station in 1978 and 1979. Information about these is given in appendix one. Nutrient deficiency symptoms produced in this way sometimes differ from the corresponding ones seen in field-grown crops: some of the possible reasons are mentioned there. Because of this, we have included photographs of field-grown crops wherever possible, even though the photographs might be inferior, and the diagnosis less certain. (If uncertainty exists, it is mentioned in the caption.) The chief source of new photographs of microelement deficiencies was a series of flowing culture experiments carried out at Long Ashton Research Station by Dr E. J. Hewitt and Mr E. F. Watson, to whom we are also indebted for much useful advice and help throughout the project.

Following the colour atlas will be found a list of contributors of other photographs, a glossary of terms and abbreviations used in the test, and an appendix contributed by Dr F. S. MacNaeidhe, of the Peatland Experimental Station, Lullymore, Ireland. His experience of micro-element disorders on Irish peat soils, communicated to us by letter, seemed likely to be of considerable interest to readers. It is therefore included very much as received from him.

We are also happy to record our grateful thanks to the very many other people who have helped in various ways with the preparation of this volume: in particular Dr G. W. Winsor and Mr P. Adams (G C R I, Littlehampton), Dr J. B. D. Robinson (L A R S, Bristol), Mr A. Barnes, Dr J. A. Tomlinson and Mr H. A. Roberts (N V R S, Wellesbourne), Mr A. J. Biddle (P G R O, Peterborough), Mr M. F. Harrod, Mr E. C. Herwin and Mr R. W. Swaine (A D A S), Prof. Dr. W. Bussler (Technical University, Berlin), Prof. S. Locascio, Prof. R. E. Lucas and Prof. D. N. Maynard (all at University of Florida), Prof. G. E. Wilcox (Purdue University, Illinois) and Mr K. W. Morris (Phosyn Chemicals Ltd, York). Mr A. K. Burton and Mr K. Moore, both sandwich students, made useful contributions to the work. The analytical figures were mainly provided by Mr J. Hunt and his staff at N V R S and although Mr Wood took most of the photographs, his successor, Mr R. Sampson, and assistant, Mrs P. Quick, have also contributed. Mrs D. Walters typed and re-typed the manuscript with remarkable patience, and many other collegues at N V R S helped in a variety of ways. We also thank the numerous people who sent photographs which were not eventually used, and our Head of Section, Dr D. J. Greenwood and Station Director, Prof. J. K. A. Bleasdale, for supporting the project.

CHAPTER 1
Diagnosing mineral disorders by eye

The diagnosis of a mineral disorder will normally proceed through five or six clearly definable steps:

1. The **realisation** that something is wrong with the crop and that the problem might be avoidable. This stumbling-block, curiously enough, is undoubtedly the prime cause of huge yield losses by less well informed farmers and gardeners throughout the world, who may either not notice that their crops are suffering, or be unaware that the symptoms they see indicate a nutrient disorder. Furthermore, for most nutrients the probable yield loss is very considerable by the time deficiency symptoms appear.

2. The **observation** in detail of all the abnormalities exhibited by the crop, and the way these are distributed on the plants and across the field. Important points to note in the case of foliar disorders are: whether they are worse on old or young leaves: whether the problem is one of chlorosis (yellowing) and whether this is largely interveinal (between veins), marginal (around leaf edges) or uniform on a given leaf: or of necrosis (death and thence drying out) of tissue, and its pattern of distribution, as for chlorosis: or of deformities such as cupping, twisting and thickening. Other symptoms to look for are cracking and brittleness, hollowness and brown water-soaked areas in stems and roots, flower and fruit abnormalities, peculiarities of plant habit, such as unusual erectness or foreshortened internodes, and spotting.

At this stage one is looking for evidence to decide whether the problem is nutritional or not. Characteristic features of nutritional disorders are that: within a small area, virtually all plants (unless genetically diverse) are likely to be affected, and on each plant all leaves of a particular physiological age are almost bound to show similar symptoms: if there is variation in the field, it is likely to follow the topography, soil type or (if a result of mis-application of fertilisers) the pattern of farming operations e.g. beds, headlands, spray-boom widths: there may be a lessening of the severity where there is reduced inter-plant competition, as on the edge of beds or beside areas of bare soil. None of these features completely rules out a non-nutritional explanation, but this type of observation may provide valuable clues. It is assumed that the reader is familiar with the commoner pests and diseases, but it is perhaps worth mentioning two characteristic effects which are never, in the writers' experience, due to nutrition. These are (a) a pattern of leaf lesions in which almost circular necrotic spots of any size from that of a pin-head upwards are surrounded by a 'halo' of chlorotic tissue (page 8 top left), and

positioned randomly in relation to the veins. Such spots are almost certainly caused by pathogens; (b) a form of chlorosis which follows the veins, leaving the interveinal areas green (page 8 top right). This is usually due to a virus, or herbicide, being carried round the vascular system and hence being most damaging in the zone nearest the veins.

Useful items to keep handy at this stage of the diagnosis are a note-book and pencil, a sharp penknife, magnifying glass, polythene bags, ties and labels to keep plant samples fresh, and a fork for digging out roots. Some method of assessing soil pH e.g. test papers, and the nitrate test strips mentioned in chapter two are also invaluable.

3. The **identification** of possible causes, by reference to the colour atlas. If any of the pictures resembles the symptoms found, one should not jump to conclusions but examine all pictures of the crop in question. If a particular diagnosis appears likely, look out for footnotes and refer back to the appropriate element in this chapter for comments about other possible causes of similar symptoms.

4. The **elimination** of unlikely explanations, using the information about the effects of soil type and pH, species susceptibility etc given in this chapter and in volume one. The quick sap test for nitrate, mentioned in chapter two, is very useful here. For example purple colours in brassica crops may be due to N or P deficiency, or cold. Use of the sap test will immediately confirm or eliminate the probability of N deficiency, provided that the crop is still relatively young.

5. If doubt remains, it will be necessary to **check** the diagnosis by means of leaf or whole-plant analysis (see chapter 2. However, our present state of knowledge about the optimum concentrations is rather imprecise, so that a firm diagnosis may still be impossible unless the concentration in the crop in question is substantially above or below the optimum. One solution to this difficulty is to compare analytical figures from 'good' and 'bad' areas of the same crop, as mentioned in chapter 2.

6. Definite **proof** of the correctness of the diagnosis is only really possible by successful application of a curative treatment (such as those mentioned later in this chapter) and comparison with untreated plants, grown alongside the treated plants and cultivated in exactly the same way. For some elements where curing the current crop is impracticable,

Alternaria blight in Brussels
sprout. This type of round necrotic
lesion, scattered randomly over the
leaf, is rarely, if ever, due to a
mineral disorder.

Residues of substituted urea and
uracil type herbicides in the soil
can cause striking 'vein-clearing'
patterns in crops which are,
however, unlike those resulting
from any mineral disorder.

it will be necessary to apply preventative treatments to future
crops. In the case of toxicities, it is rarely feasible to remove
the unwanted element from the soil, and the treatment
usually depends on adjustment of soil pH. This of course may
have effects other than the one intended, and is therefore not
strictly diagnostic. In any event, the grower is strongly urged
to leave a narrow, untreated test-strip whenever applying
treatments which may be of doubtful efficacy.

Ideally, one would like to provide an infallible routine,
similar to a botanical flora, for identifying disorders. Such
'keys' have been published for this purpose e.g. by English and
Maynard, (1978), but we have adopted the alternative procedure of listing generalised symptoms, predisposing factors,
and species susceptibilities under each element, even though
this requires the reader to absorb more information before he
can draw conclusions. This is because the 'key' approach
'switches' the diagnosis irrevocably in a particular direction
according to a single character, such as leaf colour, whereas
we feel it to be important to keep in mind all relevant factors
whilst arriving at a diagnosis. Also, for certain elements, a

generalised key for all vegetable crops is hardly practicable
because the symptoms differ from crop to crop.

Despite this, we have included generalised symptoms in the
following pages to lead the reader to the most likely diagnosis.
A more reliable syndrome description, specific to the crop,
will be found in the colour atlas. The remarks on 'occurrence
and predisposing factors' must be considered carefully to
avoid arriving at highly improbable conclusions: the ranking
under 'species susceptibility' refers only to the relative
sensitivity of crops to that particular disorder, *assuming that
conditions are marginal for its occurrence.* Thus, although
onions are said to be most sensitive to zinc deficiency, this
should not be taken to mean that they will suffer from this
disorder more often than from potassium deficiency, to which
they are only fairly susceptible. The 'critical concentrations'
(which are expressed on a dry matter basis) apply only to a
sample of 'middle' leaves, taken half-way through crop life, as
discussed in chapter two. The term 'cure' applies to the
possibility of curing the already affected crop, and implies that
growth made subsequent to the treatment should be healthy,

even if tissues already affected do not recover. Wherever foliar sprays are recommended, an appropriate wetting agent should be added.

Nitrogen deficiency

Symptoms. Fairly uniform pale colour, leading to yellowing, especially on lower leaves, of most species. Feeble growth and lack of branching. Many brassicas and other species show purple flush on leaves and/or pink midribs, petioles and stems. In all species petiole sap nitrate concentrations (see p. 18) of young plants showing deficiency symptoms are below 500 µg ml^{-1} (NO_3).

Occurrence and predisposing factors. Very common whenever insufficient N fertiliser or well-decomposed organic matter applied to crop: on soils low in organic matter: after very heavy rain and waterlogging causing leaching and/or de-nitrification: on dry soil, preventing nitrate movement to roots: with close spacing etc giving high yield potential hence high N demand: after use of insufficiently decomposed organic manure, eg strawey residues which immobilise soil mineral N: for pot plants, when not fed continuously or re-potted: where soil temperature is too low to permit mineral-isation of organic matter e.g. in spring.

Species susceptibility: all species except nodulated legumes badly affected but quantities of N fertiliser needed vary greatly as follows:
brassicas, red beet, 100–300 kg ha^{-1}N:
french and runner beans, celery, leeks, lettuce, marrows, onions, spinach and sweet corn, 50–200 kg ha^{-1}N:
broad beans, carrots, parsnips, peas, radish, swedes, and turnips, 0–100 kg ha^{-1}N. Exact amounts required depend on predisposing factors mentioned above and are difficult to quantify.

Similar symptoms result from: cold weather (purple/pink colours in brassicas): root damage e.g. by nematodes: drought and waterlogging (often for reasons given above).

Critical leaf concentration: 3.5%N.

Prevention: use of recommended rate and timing of N fertilisers e.g. for England and Wales, see M A F F (1979).

Cure: side-dressings of quick-acting nitrogenous fertiliser e.g. ammonium nitrate, watered in if no rain falls: or foliar sprays of 2% urea at high volume (500–1000 l ha^{-1}).

Phosphorus deficiency

Symptoms: reduced growth rate, particularly soon after emergence, often with no other symptom. Some species

(brassicas, carrots, sweet corn) show purpling of older leaves, and there is often a tendency towards dull, bluish green leaves. Recovery can occur.

Occurrence and predisposing factors: common wherever P fertilisers are not used, but especially on acid soils rich in iron and aluminium oxides (e.g. red tropical soils), calcareous soils, strongly adsorbing clays or peats. Aggravated by cold weather at emergence.

Species susceptibility: most susceptible: carrot, lettuce, spinach, watercress. Fairly susceptible: broad bean, broccoli, french bean, onion, sweet corn, turnip. Least susceptible: Brussels sprout, cabbage, cauliflower, leek, parsnip, pea, radish, red beet, swede (based on yield response in field experiments at N V R S).

Similar symptoms result from: low temperatures, drought, root pests e.g. carrot fly, aluminium toxicity on acid mineral soils (may actually be P deficiency).

Critical leaf concentration: 0.35% P.

Prevention: use of appropriate P fertiliser in seed-bed (M A F F 1979). On strongly 'P-fixing' soils it is essential to place the fertiliser in close proximity to the seed.

Cure: not practical. Foliar sprays phytotoxic.

Sulphur deficiency

Symptoms: new leaves uniform golden yellow, (celery, french beans, lettuce, marrow, sweet corn) or cupped and deformed, with diffuse interveinal chlorosis, (crucifers, red beet). Foliage frequently stiff and erect e.g. broad bean.

Occurrence and predisposing factors: occasionally found on low organic matter soils far from sources of industrial sulphur dioxide. Has often appeared when fertilisers with low sulphur content e.g. triple superphosphate, replaced those with high sulphur content e.g. single superphosphate. Not yet recorded in UK, but found elsewhere in Europe and all other continents.

Species susceptibility: most susceptible: cabbage, curly kale, leek, radish, swede, turnip. Fairly susceptible: broccoli, Brussels sprouts, cauliflower, lettuce, onion, pea, red beet, spinach, sweet corn. Least susceptible: broad bean, carrot, celery, french bean, parsnip, watercress (based on growth scores on N V R S pot experiment).

Similar symptoms result from: iron deficiency (yellowing of new leaves) although venation remains green initially in this case.

Critical leaf concentration: 0.2% S.

Prevention: use of sulphur-containing fertilisers such as ammonium sulphate, single superphosphate, or else apply gypsum at 100 kg ha^{-1}. Elemental sulphur is also used, although this has the effect of acidifying the soil and takes several months to have full effect.

Potassium deficiency

Symptoms: marginal scorch of older leaves in most crops. Scorched margins curl up or down. Shortened internodes in peas. Wilting and early abscission. Marginal scorch may be preceded by marginal spotting, chlorosis, or mesophyll collapse.

Occurrence and predisposing factors: would occur in intensive horticulture but for regular fertiliser potassium dressing, except on certain clay soils which release potassium steadily from lattice. Likelihood increased by low soil exchangeable K, high levels of other cations e.g. magnesium, ammonium.

Species susceptibility: most susceptible: spinach. Fairly susceptible: broad bean, broccoli, cauliflower, french bean, leek, lettuce, onion, radish, red beet, sweet corn, turnip. Least susceptible: Brussels sprout, cabbage, carrot, parsnip, pea, swede, watercress (based on yield response in field experiments at N V R S).

Similar symptoms result from: wind damage, chloride toxicity (especially where irrigation or tapwater used) but both the above tend to occur on younger leaves: thiocarbamate—based herbicides.

Critical leaf concentration: 2.0% K.

Prevent by: use of recommended rate of K fertiliser for crop (M A F F, 1979).

Cure: not practical. Foliar sprays phytotoxic.

Calcium deficiency

Symptoms: cupping (both convex and concave) and tipburn or extensive blackening of young leaves: or pale marginal band on young leaves: appearance of watersoaked areas across leaves, petioles or stems which then collapse. Gelatinisation of root tips. Several important vegetable disorders such as lettuce tipburn, internal browning of brussels sprouts, and blackheart of celery are associated with localised calcium deficiency.

Occurrence and predisposing factors: occurrence is erratic and unpredictable, and can affect 100% of the crop in the case of lettuce and celery. Associated with rapid growth in hot weather, and dry soil. Internal browning of Brussels sprouts is rarer in U K but a small incidence (1–2%) can result in crop rejection by processors. Absolute lack of soil calcium rarely likely to be the cause, although liming to pH 6.5–7.0 is wise. More often due to antagonism from other cations e.g. magnesium, potassium, ammonium. Genetic resistance exists e.g. in 'Salinas' lettuce, but there are large interactions with season and site.

Species susceptibility (in commercial practice): most susceptible: celery, lettuce. Fairly susceptible: Brussels sprout, cabbage, sweet corn. Other species not regularly affected.

Similar symptoms result from: frost damage (marginal leaf scorch), benzonitrile herbicides (kill growing point): thiocarbamate herbicides (cause leaves to cup and stick together).

Critical leaf concentration: whole leaf analysis for Ca is fairly meaningless: even for localised, susceptible tissues the critical concentration appears to be very variable.

Prevention: soil should be limed normally to recommended pH (M A F F, 1979) but this does not guarantee freedom from calcium-related disorders. Avoid excessive soluble salts and soil compaction. Irrigate vegetables regularly. For control of blackheart in celery, fortnightly sprays of Ca(NO$_3$)$_2$ (16 kg 1000 l^{-1} ha^{-1}) or CaCl$_2$ (8 kg 1000 l^{-1} ha^{-1}) are effective. Resistant varieties sometimes obtainable.

Cure: impossible for vegetable disorders in field crops.

Magnesium deficiency

Symptoms: interveinal chlorosis beginning on older leaves, giving a mottled appearance. The primary and secondary veins, but not the very finest (tertiary) veins normally remain green. Very characteristically the leaf margin sometimes remains green, notably in legumes. Brassica leaves may develop brilliant orange, yellow and purple colours, especially on underside of leaves. In extreme cases the chlorotic tissue dies, leaving brown areas. Badly affected leaves may absciss.

Occurrence and predisposing factors: fairly common, especially as crop maturity approaches, on sandy soils and in wet weather. When it appears towards maturity yield is normally not affected, although the appearance may reduce marketability. Soil compaction, waterlogging and water stress aggravate disorder.

Species susceptibility: most susceptible: cauliflower, lettuce. Fairly susceptible: broccoli, cabbage, carrot, curly kale, marrow, onion, pea, radish. Least susceptible: broad bean, Brussels sprout, celery, french beans, leek, parsnip, red beet, spinach, swede, sweet corn, turnip, watercress (except for watercress, based on growth scores on N V R S pot experiment).

Similar symptoms result from: virus infection (see footnotes to individual species in colour atlas).

Critical leaf concentration: 0.2% Mg.

Prevention: incorporation of magnesian limestone, (9–11% Mg), Epsom salts (10% Mg), Kieserite (16% Mg), Kainit (3% Mg) or other Mg-containing fertilisers into the soil to provide recommended amounts of Mg for crop (M A F F 1979): or fortnightly foliar sprays of 2% Epsom salts ($MgSO_4$ $7H_2O$) at high volume (500–1000 l ha^{-1}).

Cure: sprays as above.

Iron deficiency

Symptoms: interveinal or uniform chlorosis of young leaves, which become progressively yellower or whiter as the deficiency becomes more serious. No deformity is present, but brown necrotic spots may occur on the bleached foliage. In the very youngest leaves all trace of green may be absent, leaving a strikingly white leaf.

Occurrence and predisposing factors: rare in vegetable crops in Britain although not uncommon on fruit crops. Associated with calcareous soils. Likely to occur in hydroponic systems especially where no solid substrate exists to retain insoluble iron compounds in contact with roots. Also induced by waterlogging, cold soil, and excess heavy metals, such as copper, cadmium, nickel and zinc (Hewitt, 1948). Excess P may also induce it.

Species susceptibility: most susceptible: sweet corn, watercress. Fairly susceptible: brassicas, french bean, pea, red beet. Others not susceptible. Clear-cut genetic differences in susceptibility occur within species.

Similar symptoms result from: manganese deficiency, which may occur with it. Certain herbicides such as paraquat may produce a similar bleaching of new leaves. Proximity to certain types of plastic eg flexible P V C.

Critical leaf concentration: about 50 µg g^{-1} Fe, but not very reliable. Appropriate preparation of samples imperative (see volume 1).

Prevention: soil applications of iron chelates, preferably E D D H A ('fritted' iron is used successfully in other countries), or fortnightly foliar sprays of 0.05% Fe E D T A at high volume (500–1000 l ha^{-1}).

Cure: sprays as above.

Manganese deficiency

Symptoms: interveinal chlorosis which begins on new leaves but may soon spread to others. Furthermore, the plant may 'grow away' from the disorder so that new leaves appear less affected than older ones. The green colour remains in the very finest veins giving an intensely reticulated pattern and the chlorotic areas may be bleached white. Seeds of peas and beans show 'marsh spot' when split open (pp. 34, 61, 65). Leaf analysis valuable for confirmation.

Occurrence and predisposing factors: fairly common on peaty or heathland soils above pH 6.5, and on calcareous mineral soils with poor drainage. Sometimes seen on tractor wheelings.

Species susceptibility: most susceptible: french bean, onion, pea. Least susceptible: leek. Rest fairly susceptible.

Similar symptoms result from: iron deficiency: magnesium deficiency (though latter never found on youngest leaves).

Critical leaf concentration: 20 µg g^{-1} Mn.

Prevention: avoid overliming: Mn application to soil not recommended. One or more foliar sprays of manganese sulphate at 9 kg ha^{-1}. Peas — 5 kg ha^{-1} at early flowering stage.

Cure: spray as above.

Zinc deficiency

Symptoms: see individual crops.

Occurrence and predisposing factors: not recorded in UK vegetable crops but fairly common in USA (see reference below), especially on soils of high pH or high P content, and in cool wet weather. Serious following land-levelling operations where topsoil removed, exposing calcareous subsoil.

Species susceptibility: most susceptible: Phaseolus beans, onions, sweet corn. Moderately susceptible: cabbage, carrots, celery, lettuce, peas, spinach.

Critical leaf concentration: 20 µg g^{-1} Zn.

Prevention: use zinc-supplemented fertilisers to supply 2–20 kg ha^{-1} Zn, according to local recommendations.

Cure: spray zinc sulphate ($ZnSO_4$ $7H_2O$) at 5 kg 500 l^{-1} ha^{-1}: (ineffective for onions). See 'Zinc in crop nutrition' International Lead Zinc Research Organisation, for further details).

Copper deficiency

Symptoms: very dependent on species, and several e.g. carrots, suffer yield loss without symptoms appearing. Generally, the new leaves become greyish-green, chlorotic, or even white, with some wilting, and growth is severely reduced.

Occurrence and predisposing factors: only occurs on peat soils or on organic chalky or heathland soils, which have received no recent copper application. Very severe on newly reclaimed peats. (See appendix 2 and Pizer et al., 1966).

Species susceptibility: most susceptible: carrot, lettuce, onion. Fairly susceptible: broad bean, brassicas, celery, pea, radish, red beet, spinach, sweet corn, turnip.

Critical leaf concentration: 5 μg g^{-1} Cu.

Prevention: soil applications of copper sulphate (CuSO$_4$ 5H$_2$O) at 60 kg ha^{-1} at roughly 10-year intervals, or foliar sprays of copper oxychloride (CuOCl) or cuprous oxide Cu$_2$O at 2 kg ha^{-1}, low or high volume.

Cure: sprays as above.

Boron deficiency

Symptoms: brittle tissues which crack easily: hence regularly spaced transverse cracks across petioles: cracked, corky areas on stems and midribs and split storage roots. Death of growing point and distortion and blackening of new leaves, with consequent loss of apical dominance and outgrowth of side shoots. Hollow stems and interveinal chlorosis in brassicas, 'brown heart' of swede and turnip roots, and cankerous black spots in red beet roots.

Occurrence and predisposing factors: fairly common in areas of high rainfall especially on light soils at high pH (over 6.5). Symptoms often occur soon after the end of a drought, see Appendix 2. Aggravated by high N use, recent liming, and by dry summer following wet winter (MAFF, 1976). For a recent review on all aspects of B deficiency see Gupta (1979).

Species susceptibility: most susceptible: cauliflower, celery, red beet, spinach, swede. Fairly susceptible: Brussels sprout, carrot, leek, lettuce, marrow, radish.

Similar symptoms result from: stem eelworm damage on root crops (produce cankerous areas on red beet roots), benzonitrile herbicides (kill growing point): growth regulator herbicides (cracking and corkiness of stems, especially brassicas, see plate above right).

Critical leaf concentration: 20 μg g^{-1} B.

Severe splitting and corkiness of Brussels sprout stem caused by 2–4–5T drift. When splitting is less pronounced, this can closely resemble boron deficiency (see Plate 34).

Prevention: apply borax or other soluble borate at 2 kg ha^{-1} B (20 kg ha^{-1} borax) before sowing, or spray 'Solubor' at 10 kg ha^{-1} at early growth stage.

Cure: spray as above.

Molybdenum deficiency

Symptoms: 'Whiptail' of cauliflower—lamina of new leaves progressively twisted and reduced, until mibrib only appears. Growing point becomes 'blind'. Young brassicas show cupping and interveinal necrosis.

Occurrence and predisposing factors: rare, except in cauliflowers (other species often affected in Australia and New Zealand). Unlike most other trace element deficiencies that of molybdenum is most likely on acid soils (below pH 5.5), but it can occur in cauliflower even at pH 7.0. Below pH 4.5, it is likely to be masked by toxicities of manganese or aluminium.

Species susceptibility: most susceptible: cauliflower, spinach. Fairly susceptible: Brussels sprout, cabbage, lettuce, onion, red beet, turnip. Slightly susceptible: broccoli, carrot, parsnip. Least susceptible: pea, Phaseolus beans, radish, sweet corn. (Main source, Johnson et al., 1952).

Similar symptoms result from: swede midge attack in cauliflower which causes distortion of young leaves and blindness —larvae should be present. Sudden cold can induce abrupt blindness, without progressive reduction of leaf lamina. See p. 31.

Critical leaf concentration: 0.1 $\mu g\ g^{-1}$ Mo.

Prevention: lime soil to pH 6.5. Also (for cauliflowers only) apply sodium or ammonium molybdate solution (0.03%) as foliar spray or soil drench at 1000 l ha^{-1} at an early growth stage, or treat seed with 35 g ha^{-1} molybdate (U S A method).

Cure: impossible once whiptail present.

Chlorine deficiency

Symptoms: wilting: restricted and highly branched root system, often with stubby tips.

Occurrence and predisposing factors: unknown in field-grown crops, although amounts required for growth are higher than for any other micro-element and chloride is readily leached. Plants appear to have considerable ability to accumulate chlorine from the atmosphere (Johnson et al., 1957). These authors point out that deposition in rainfall usually exceeds crop requirement by a factor of at least five.

Species susceptibility: most susceptible: lettuce, cabbage, carrot. Fairly susceptible: bean. Least susceptible: squash. (Johnson et al., 1957).

Similar symptoms result from: anything causing wilting.

Critical concentration in leaves: 100 $\mu g\ g^{-1}$ Cl.

Prevention and cure: unnecessary—even in hydroponic systems sufficient should be present as impurities in chemicals etc.

Sulphur toxicity

This occurs in two ways: as a build-up of soil sulphate from irrigation water and as a result of gaseous sulphur dioxide in the atmosphere. Neither are within the scope of this book, but

it may be mentioned that the symptoms of sulphur dioxide toxicity are either acute, in which case there is tissue collapse round leaf margins or between veins, and the collapsed tissue then bleaches to an ivory or brown colour (depending on species); or chronic,—'a yellowing of leaves which progresses slowly through stages of bleaching till . . . interveinal areas of leaf are nearly white' (Thomas, 1951). In experiments with cauliflower at NVRS, the reaction closely resembled tissue collapse caused by calcium deficiency (Plate 17). (Hardwick, R. C., personal communication).

Manganese toxicity

Symptoms: cupping, interveinal chlorosis and necrotic spotting of leaves in brassicas. Black spotting on petioles and veins on Phaseolus beans. On some soils, iron deficiency may be induced. Normally occurs as a result of soil acidity in which case symptoms of calcium and molybdenum deficiency may also be present, as a 'soil acidity complex'.

Occurrence and predisposing factors: susceptible species very likely to suffer if soil pH falls below 5.0.

Species susceptibility: most susceptible: Brassicas and Phaseolus beans. Fairly susceptible: swedes, cauliflowers, lettuce. (Hewitt and Smith, 1975).

Upper critical leaf concentration: (See chapter 2, also volume 1): 500 $\mu g\ g^{-1}$ Mn — but varies with species.

Prevention: lime soil to pH 6.5.

Cure: not possible.

Boron toxicity

Symptoms: marginal chlorotic band on old leaves.

Occurrence and predisposing factors: occasionally found as a result of application of too much B as fertiliser (most crops are very sensitive). Low pH.

Species susceptibility: most susceptible: Phaseolus beans.

Upper critical leaf concentration: very variable—between 100 and 500 $\mu g\ g^{-1}$ B.

Prevention: when applying boron, ensure very exact application rate and even distribution e.g. by dissolving in water and spraying onto soil.

Aluminium toxicity

Symptoms: normally as for P deficiency—ie slow growth, dull purple leaves: but see individual species e.g. beans, p. 67. Roots discoloured before tops injured. Generally symptoms not very diagnostic.

Occurrence and predisposing factors: susceptible species likely to suffer on acid mineral soils (pH below 5.0) containing aluminium compounds (See Chapman, 1966). Condition is aggravated by low P status.

Species susceptibility: most susceptible: celery, beet, lettuce. Fairly susceptible: cabbage, radish. Least susceptible: maize (therefore sweet corn?), turnip. (Chapman, 1966, Hewitt and Smith, 1975).

Upper critical leaf concentration: not considered useful.

Prevention: lime to pH 6.5. Use generous P dressings.

Cure: impractical.

Chlorine toxicity

Symptoms: marginal leaf scorch, abscission, chlorosis.

Occurrence and predisposing factors: usually associated with irrigating with chloride-containing water: in such circumstances the disorder is fairly common.

Species susceptibility: most susceptible: beans, peas, onions, lettuce. Fairly susceptible: carrot, leek, spinach. Least susceptible: red beet, brassicas.

Upper critical leaf concentration: symptoms likely to appear above about 0.5% Cl for sensitive crops; up to 4% for tolerant species.

Prevention and cure: by leaching out excess salts.

Plant analysis and sap testing

This has been dealt with in detail by Bould in volume 1. The purpose here is to provide summarised notes on the collection of leaf samples intended for laboratory analysis, some detail about quick sap tests used at N V R S and guidance on the interpretation of results from both methods for vegetable crops.

For the benefit of readers who have not read volume one, chapter three, it is first necessary to explain what is meant by the term 'critical concentration'. For vegetables, the possibility exists of quality as well as yield being affected by a disorder, and we would therefore define the critical concentration of an element in a sample of leaves taken early in crop life as 'that which is just deficient for eventual maximum *marketable* yield'. An 'upper critical concentration' also exists which is 'just excessive' for eventual maximum marketable yield. Between the two critical concentrations there is no effect on yield. The first sign of symptoms does not normally correspond with the attainment of either of these concentrations.

There are two likely reasons for resorting to plant analysis or sap testing. The first is to confirm a diagnosis based on the appearance of symptoms: in such cases the concentrations will usually be well below the critical level and there should therefore be little doubt about the diagnosis. Typical analysis figures for such situations are given in the picture captions wherever they are available. The other reason is to test for subclinical deficiencies or toxicities which are already limiting growth but which are not yet resulting in visible symptoms. For this purpose much more care is required in the selection of leaves and in making allowance for plant or leaf age when comparing results with established critical levels. The most important factors here are (a) to sample a sufficiently large number of plants, (b) to take leaves from the position on the plant and at the growth stage for which 'critical values' are available for comparison and (c) to use analytical or sap testing methods having the same accuracy (percentage recovery) as those used in establishing the critical concentrations. This should normally approach 100%.

In view of the large errors associated with established critical concentrations, it is wise whenever possible to allow the affected crop to provide its own 'control' values if there are healthy, fast-growing areas within it which can be used for this purpose. When this is the case a sample of leaves from the healthy crop should be collected and analysed in exactly the same way as the unhealthy area. Provided that the samples are then analysed for all 'suspect' elements, the one responsible for the disorder should show a substantial difference between good and bad areas of the field.

Leaf sampling for laboratory analysis

To adequately represent any field or smaller area of crop, at least 25 plants should be sampled. The A D A S method is to walk through the crop following a 'W' pattern, collecting leaves at regular intervals. For most elements, a single undamaged, 'mature physiologically active' leaf is taken from each plant, without any attached petiole. (In the case of compound leaves such as those of carrots, french beans, etc we discard the main petiole but include the petioles of the individual leaflets in the sample). The selection of the 'right' leaf can be fraught with some difficulty, and we have come to the conclusion that for most crops the most practical procedure is to count the leaves greater than 1 cm long, excluding scars of abscissed leaves, and take the middle one of these. This clearly takes some time for the first plant, but thereafter, having noted the position of this leaf on the plant fairly carefully, one collects the remaining leaves from a similar position by guesswork. For hearted lettuce etc it is more practical to take the outermost wrapper leaf of the heart. If at all contaminated eg by sprays, the leaves should be washed as described on p. 116 of volume one, which also gives the drying and grinding procedure. Normally the fresh leaves will be sent straight to the analytical laboratory.

For analysis for the non-mobile elements sulphur, calcium, copper and boron, young leaves should be selected for analysis, since they are the most affected by the deficiency.

Some authors have sampled the entire above-ground part of the plant. This is not convenient for large plants, but it has been done on N P K experiments at N V R S and these have provided the basis of the critical concentrations given in Fig. 1. The background to this figure is given by Greenwood et al., 1980.

The selection of a 'mature physiologically active' leaf does not mean that the critical concentration remains constant throughout crop life, judging by observations made at N V R S. We found that the total N concentration in brussels sprout leaves sampled in this manner varied from 5.9% in late June to 2.9% in early October, after which it rose slightly. This was on plots which almost certainly received an optimal nitrogen supply throughout growth. Similar trends occur with phosphorus and potassium, whereas calcium and magnesium concentrations tend to increase with plant age.

Tables of leaf nutrient concentrations found in apparently healthy crops appear in Wallace (1961), Chapman (1966), Geraldson et al. (1973) and Needham (1976). Relatively few of the figures quoted for vegetables are claimed to be the critical value, however, and for each species it is possible, in many

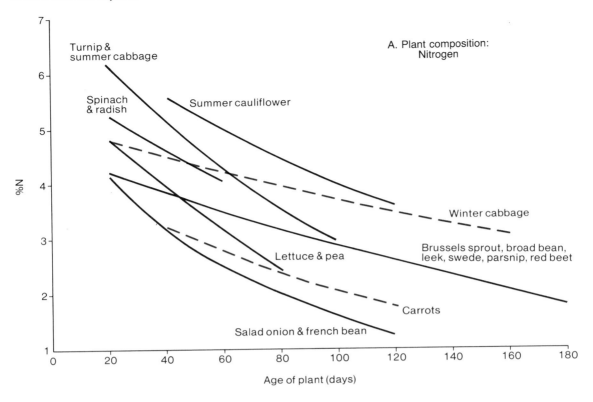

A. Plant composition: Nitrogen

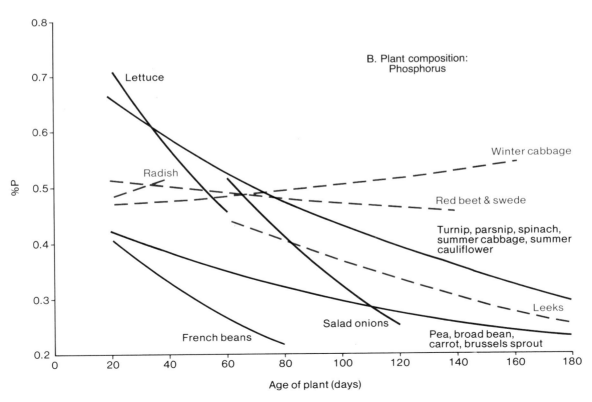

B. Plant composition: Phosphorus

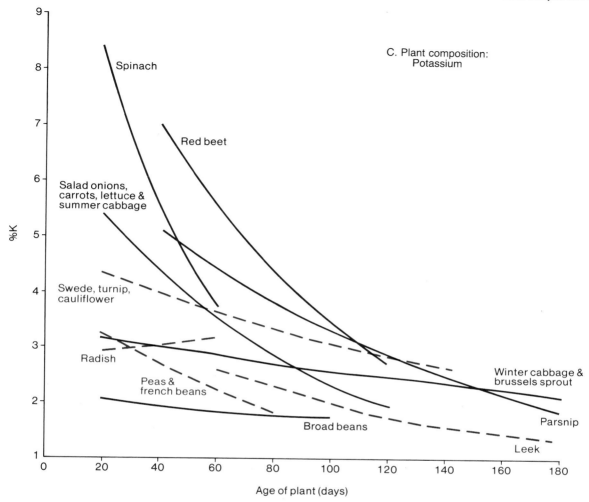

Figure 1 A, B, C Critical concentrations of N, P and K in whole
plants as a function of plant age (derived from Greenwood et al., 1980)

cases, that the critical value is lower, or even perhaps higher, than the figures given. Since it is the critical value which is of prime interest we do not feel justified in retabulating these figures: instead we have abstracted a single figure for each element which is near the lower end of the 'healthy' range for the majority of vegetable crops sampled at midgrowth, and used this value to provide the 'critical concentration' given in chapter one. Obviously these figures represent an extremely crude estimate of the critical concentration, and any measured values falling within 10%, or even 20% of the value shown must be regarded as being inconclusive diagnostically. Indeed, it seems unlikely that an exact critical figure applicable to a particular stage of growth will ever be established for a given species, if only because plants which are on the borderline of deficiency at the time of sampling (say midgrowth) may either recover or deteriorate from then until harvest, depending on the ability of the soil to provide a steady supply of the nutrient.

Quick sap tests for field use

Sap testing using Merckoquant test strips. These strips are manufactured by E. Merck of Darmstadt, West Germany. The nitrate and calcium strips have been found particularly useful for testing plants: the potassium strips cover the required concentration range but are easily discoloured by chlorophyll. The technique used at NVRS is as follows:

1. A 'middle' leaf (see p. 15) and attached petiole are broken off. If the petiole is rather dry or woody it is advisable to trim the end off square with a sharp knife.

2. The test strip is placed on a hard flat surface with the sensitised paper square facing upwards, and the leaf held against it so that the butt end of the petiole just touches the

edge of the paper square. (In the case of the nitrate strips there are two squares: only the one nearest to the end of the strip responds to nitrate.)

3.   Using a ball-pen or pencil as a roller, the sap is driven out of the petiole onto the paper square.

4.   For the nitrate strips, one simply waits two minutes and then compares the colour attained with the standards on the tube. Frequently the colour of the highest standard on the tube (500 µg ml^{-1} NO$_3$) will be reached before two minutes have elapsed. If so, a watch should be used to measure the time taken, in seconds, for this darkest colour standard to be reached. Table 1 gives the concentrations of NO$_3$ or NO$_3$-N corresponding to various times. This relationship is linear on a log-log basis and enables the conversion from seconds to µg ml^{-1} NO$_3$ to be expressed thus:

$$NO_3 = 10^{(4.9 - 1.085 \log t)}$$

where t is the time in seconds to reach the 500 µg ml^{-1} NO$_3$ colour. (Note that the tubes are marked in mg/1 (ppm) NO$_3$: for this reason we have quoted figures throughout this volume in these units, although in volume 1 the more usual units of NO$_3$-N are used, which are roughly one quarter of the NO$_3$ values. Note also that µg ml^{-1}, mg/l and ppm are identical).

Time (seconds)	µg ml^{-1} NO$_3$	µg ml^{-1} NO$_3$-N
7.5	9000	2000
9	7000	1600
15	4000	900
20	3000	680
30	2000	450
40	1400	320
60	900	200
90	600	140

Table 1   Effect of nitrate concentration on time taken for colour to reach that of the 500 µg ml^{-1} NO$_3$ standard, for 'Merckoquant' nitrate test strips

5.   It is essential to sample leaves from at least ten randomly scattered plants in any field or area to be represented, and to find the mean result. If the plant to plant variation is very large, and the mean result is borderline for a decision about nitrogen top-dressing, a further ten plants should be sampled.

6.   Time of day can have an influence on nitrate levels, which may occasionally show a minimum value at mid-afternoon on sunny days. We try to standardise sampling times between 10 am and 12 noon, but this is not always practicable.

7.   The critical concentration is not at all constant but rises to a peak soon after crop emergence and falls very rapidly during the so-called 'grand period of growth'. Therefore it is essential, when monitoring petiole nitrate concentrations, to relate the concentration found to a very well-defined growth

stage. The critical values for each growth stage of each vegetable species are not yet well established: for the petiole of the middle leaf of non-leguminous vegetable crops, the following guide represents a rough approximation: more precise figures for individual crops are available from the NVRS.

	NO$_3$ (µg ml^{-1})	NO$_3$-N (µg ml^{-1})	Seconds to reach 500 µg ml^{-1} NO$_3$ colour
First quarter of life, once fully established:	4000	900	15
Second quarter of life:	3000	700	20
Third quarter of life:	200	50	—
Last quarter of life:	0	0	—

(Legumes, if nodulated, can grow well without nitrate in the sap)

8.   If levels lower than this are found, top-dressing with a quick-acting nitrogen fertiliser should be considered. In dry weather, however, it is possible that there is adequate mineral N in the soil but it cannot reach the roots because it needs water to do so. If this is a possibility, a number of plants should be marked, thoroughly watered and retested 48 hours later. If the petiole nitrate level rises substantially as a result of watering, the crop obviously requires water rather than nitrogen fertiliser.

9.   The calcium test strips require to be 'developed' in a hydrogen peroxide solution for 45 seconds after wetting with sap. As calcium does not move readily from old to young leaves in time of deficiency like nitrate, it is advisable to monitor the petioles of the youngest leaves in this case. Critical sap calcium levels are not known: however, the test can prove useful in the study of calcium-related physiological disorders (p. 10).

10.   For precise work with these test strips it is advisable to check the calibration against standard solutions before starting.

Use of wet reagents. A useful test for potassium is provided by a 1% aqueous solution of sodium tetraphenyl boron. A single drop of this is mixed with a single drop of sap on a black tile, using a clean glass rod. A white turbidity is produced which is compared with standards. The critical concentration appears to be in the region of 2,000 µg ml^{-1} K, above which the density of the white precipitate more or less ceases to increase with increasing K concentration. A convenient small press for extracting single drops of fairly constant volume from petioles is shown in a paper by Scaife and Bray (1977) which also explains the advantages of quick sap tests, as opposed to conventional leaf analysis.

Colour Atlas

Note that the terms used are explained in the Glossary, p. 90.

Leaf numbers are counted from the bottom of the plant, excluding cotyledons, unless stated otherwise.

For details of 'NVRS experiments' see Appendix 1.

Brassicas

1

Most common brassica crops belong to a single species, *Brassica oleracea,* and therefore it is not surprising that their reactions to nutrient imbalance are very similar. One difference is in the intensity of the red and purple pigmentation which can occur with nitrogen, phosphorus, magnesium and boron deficiencies. However this varies considerably even within a particular crop: it is also affected by the prevailing temperature. It was therefore felt to be misleading to place too much emphasis on the extent of these colours. The following descriptions should be taken to apply to cabbage, Brussels sprout, cauliflower, purple sprouting broccoli, curly kale (borecole), turnip and swede, with certain exceptions which will be mentioned. The last two crops named do not belong to the species *B. oleracea,* but nevertheless can be included here for descriptive purposes.

Healthy plant

1 Six week old cabbage. (Sand culture)

Nitrogen deficiency

Leaves pale green, becoming yellow, bronzed, pink or purple as they age, and abscissing in extreme cases. Drastic reduction of growth. In curly kale, turnip, cauliflower and young swedes the predominant colour change is to yellow-orange. In cabbage, purple sprouting broccoli, Brussels sprouts and older swedes, this also occurs but is often preceded by a muddy-purple flush on the older leaves (Plate **5**). In broccoli, cabbage, Brussels sprout and cauliflower the new leaves have a particularly grey, stiff, waxy appearance.

2

2 18 day old Brussels sprout seedling showing grey-green waxy new leaf and bronzed cotyledons. (Sand culture)

21

3

3 Same plant (left) beside healthy plant of same age to show degree of stunting associated with above symptoms. Healthy plant contained 20,000 μg ml^{-1} NO$_3$ in petiole sap, whereas N-deficient plant contained 30 μg ml^{-1}. Total N (whole plant) for healthy plant was 3.5%; deficient 1.9%. In other species (see above) similar growth and sap nitrate differences were found at this growth stage when yellow cotyledons were only obvious deficiency symptom.

4 Leaves from deficient cauliflower seedlings grown in peat compost.

5 N-deficient cabbage (cv 'Market Victor') growing in field at N V R S, showing overall pale colour and purple tinge on old leaves.

4

5

6

7

6 Left, middle leaf detached from plant shown in Plate **5** and right, corresponding leaf from healthy plant, to show the difference in green colour. Nitrate test strips (shown) indicated 0 and 20,000 $\mu g\ ml^{-1}\ NO_3$ respectively.

7 Ten week old turnip showing yellowing of old leaves and pink petioles. Mean total N in leaves 1 and 2 was 1.4%. Healthy plant (same leaves) 3.7%.

Note: Sudden cold and P deficiency produce purple tints: confirm N deficiency with nitrate test strips.

Phosphorus deficiency

Muddy purple flush on old leaves, as for nitrogen deficiency but without yellow-orange colours. Drastic reduction in growth associated with scarcely noticeable symptoms. In broccoli and cauliflowers leaves may be abnormally stiff and erect. Cauliflower may have red curd.

8 Dull purple flush on leaf of four week old swede plant. (Sand culture)

9 Red curd and stiff, pointed, purple-edged leaves on cauliflower.

Note: See footnote on nitrogen deficiency.

8

9

10

Sulphur deficiency

Young leaves show interveinal chlorosis, cup both concavely (mainly while emerging) and convexly, and eventually fail to grow. The chlorosis is very characteristic on *B. oleracea* types, in that the primary and secondary veins stand out as a rather blurred, blue-green pattern against a pale green background. On the underside of the leaf these dark areas are purple, and this purple or bronze coloration may later affect whole leaves: or necrosis of leaf tips spreads inward until most leaves desiccate and fall off.

10 Three week old Brussels sprout plant showing concave cupping and twisting of new leaves, and incipient chlorosis. (Sand culture)

11 Four week old purple sprouting broccoli: top left, first leaf; top right, second leaf; bottom left, third leaf (0.10% S); bottom right, fourth leaf with characteristic chlorosis (0.11% S). (Sand culture)

12 Three week old turnip, showing chlorosis and necrosis moving in from leaf tip. (Sand culture)

Note: Distinguished from magnesium deficiency in that chlorosis is on young leaves, and accompanied by cupping and distortion of these leaves. For a description of a field occurrence in green broccoli (calabrese) in Washington State, USA, see Tomkins et al., (1965).

11

12

13

Potassium deficiency

Marginal and interveinal scorch of old leaves. Scorched edges curl upwards. Scorching apparently results from tissue collapse, which is frequently observed to precede it. Some blotchy chlorosis may also be present on old leaves. Leaves may arch backwards lengthwise.

13 Two scorched leaves from deficient cabbage. The right hand one shows extreme incurling of margin.

14

15

17

14 First leaf of three week old turnip, showing tissue collapse and scorch confined to tip of leaf. K concentration 0.25%; healthy leaf 2.2%. (Sand culture)

Calcium deficiency

Cupping, distortion e.g. 'beaking' and tipburn of young leaves, which may lead to death of growing point. Frequently it appears as though a string had been threaded around the leaf edge and drawn tight, causing the leaf to crinkle and cup deeply (as in Brussels sprout, Plate **18**), or buckle upwards, so that the 'cup' begins to turn itself inside out (plate **15**). Such effects are seen in field crops.

In sand culture, another common symptom is tissue collapse which occurs in bands across leaves, so that the distal part of the leaf hangs down and eventually shrivels up. The collapsed zone may also occur at the leaf edge. In turnip, the collapse begins with yellowing.

In Brussels sprout, a commercially important physiological disorder known as 'internal browning' has been induced experimentally by withholding calcium (Millikan and Hanger, 1966). Similarly, internal tipburn of stored cabbage was attributed to calcium deficiency by Walker and Edgington (1957).

15 Typical leaf distortion in five week old cauliflower. (Sand culture)

16 Deformities and necrosis in young leaves of 14 week old cauliflower. (At this stage two mature leaves contained 0.11% Ca.) 'A' indicates typical 'beak' and 'B' points to an area of tissue collapse. (Sand culture)

17 Tissue collapse in five week old cabbage. Note collapsed area is near base of leaf in young leaf 'A' and round margin on older leaf 'B'. Mean Ca% of leaves 4 and 5 was 0.057; healthy 2.2. (Sand culture)

18 Cupping, crinkling and tipburn of young Brussels sprout leaves on Ca-deficient plant, left. Healthy plant on right.

18

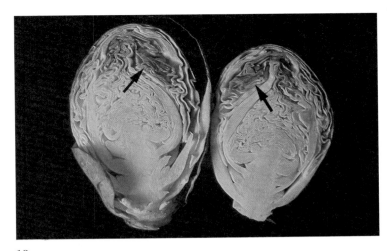

19

20

22

21

23

(Calcium deficiency, cont'd)

19 'Internal browning' (arrowed). (Field occurrence)

Note: Tissue collapse as in Illustration 17 can occur as a necrotic reaction to turnip mosaic virus (Tomlinson and Ward, 1978), or as a result of sulphur dioxide toxicity.

Magnesium deficiency

Interveinal chlorosis of old leaves, beginning faint and diffuse but eventually becoming very striking. The chlorosis may be accompanied by a blotchy purple colour, especially between veins, near margins and on underside of leaf. Towards maturity other brilliant orange and red tints may occur.

20 Young cabbage showing early symptoms. Left, upper surface of second leaf containing 0.029% Mg; and right, under surface of third leaf, containing 0.035% Mg, to show purpling on latter. Healthy plants: second leaf, 0.46% Mg, third, 0.40% Mg. (Sand culture)

21 Typical advanced chlorosis on old leaf of field-grown cauliflower. Mg concentration, 0.033%.

22 Some swede varieties exhibit chlorosis as in Plate **21**. Others, such as 'Western Perfection', shown here, show a blotchy reddening moving in from the old leaf edges. Left, fifth leaf; right, third leaf from four week old plant. Mg concentration in the two combined leaves, 0.018%; healthy plant 0.25%. (Sand culture)

23 In turnip, the leaf margins may initially remain green, as here. Mg concentration 0.061%; healthy 0.44%. Later, whole leaf affected. (Sand culture)

24 Vein clearing in cauliflower caused by cauliflower mosaic virus, see footnote.

Note: Some resemblance to certain reactions to cauliflower mosaic virus (Shepherd, 1970). 'Vein clearing' in young leaves, as in Plate **24** is a sign that the virus is present. In turnip, beet western yellows virus resembles the deficiency (Duffus, 1972).

24

Iron deficiency

Young leaves turn yellow and later virtually white, with some green colour remaining along midrib and main veins. In swede and turnip chlorotic mottling of all foliage. Unlikely except in hydroponic systems or as a result of heavy metal toxicity.

25 Cauliflower: iron deficient plant on left, healthy plant on right. (Flowing solution culture)

25

Manganese deficiency

Interveinal chlorosis of all leaves, presenting a 'freckled' or 'speckled' appearance. The chlorotic areas have an olive green colour rather than yellow as in magnesium deficiency, with which it is often confused. Manganese deficiency may also resemble iron deficiency as Plate **27** shows. Leaf analysis needed to confirm.

26 14 week old cabbage, showing overall speckled chlorosis. Young leaves contained $1.3\ \mu g\ g^{-1}$ Mn (Peat culture)

27 16 week old purple sprouting broccoli plant showing bleaching of young leaves, very much as in iron deficiency. Pink tinges on petioles, veins and leaf edges. (Peat culture)

26

27

28

29

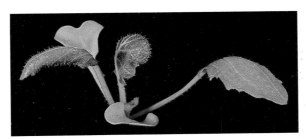

30

31

Zinc deficiency

Cabbage: pitcher-like cupping with outcurved margins, of expanding leaves. Interveinal bronzing of old leaves.

28 Left, zinc deficient cabbage; right, healthy plant. (Flowing solution culture)

Copper deficiency

Cabbage: feint diffuse interveinal chlorosis of expanding and mature leaves. Oldest leaves withered. Very rare.

29 Young deficient cabbage plant grown in flowing solution culture, showing chlorosis on younger leaves and collapse of older ones.

Boron deficiency

Brassica crops are sensitive to boron deficiency and exhibit numerous very characteristic symptoms, but not all of these occur on all species. The relevant crop(s) will therefore be mentioned for each symptom.

If the deficiency occurs at the seedling stage, the new leaves show convex cupping. (Turnip, cabbage, cauliflower only—see Plate **30**). In the rather extreme situation of the NVRS peat experiment, cauliflower reacted by producing thick, brittle, finger-like new leaves, and the cotyledons grew very large (Plate **31**).

Later on, interveinal chlorosis, worst on old leaves, (which have red margins) occurs on Brussels sprouts (Plate **32**). On broccoli, swede, and turnip the chlorosis is marginal, with brilliant red and yellow colours (Plate **33**).

Cracked and corky stems (Plate **34**) petioles, and midribs (Plate **35**) occur on all species except possibly turnips.

The growing point does not normally die in field crops, but some of the consequences of its moribund condition may appear, eg multiple crowns (swede and turnip) and twinning of stems (Brussels sprouts). In the NVRS experiment cabbages rosetted (Plate **36**) and newest leaves were hooked and scorched at edges (Plate **37**). In the field, cabbage

32

35

33

36

34

37

heads may simply be small and yellow.

The two root crops exhibit 'brown heart' (also called 'raan' or 'water core', see Plate **38**) in which the root tissues are brown and watersoaked, often with one or more hollow cavities near the crown. Later, the root becomes increasingly hollow, fibrous, bitter and invaded by rotting organisms (Plate **39**). The outside may meanwhile take on a rough, corky, leathery appearance (Plate **41**). These root symptoms occur in field crops without any other symptoms necessarily being present on the tops.

Hollows also are found in the stems of cabbage, Brussels sprouts and cauliflower (Plate **40**) although this disorder does not always respond to treatment with boron. (Gupta and Cutcliffe, 1973). Sometimes the hollows are smooth and white inside: at others they are lined with brown, evil-smelling bacterial slime.

Cauliflowers also show browning of the curd (Plate **42**) and Brussels sprouts will bear few sprouts if the deficiency sets in before they are formed: if it occurs later, the sprouts are small and loose. Shorrocks (1974) mentions rolling and curling of leaves of both species.

30 Seedling turnip showing convex cupping and dark colour of new leaves. (Peat culture)

31 Cauliflower seedling showing deformed, thick finger-like new leaves. (Peat culture)

32 14 week old Brussels sprout plant showing misshapen leaves with several forms of interveinal chlorosis. (Peat culture)

33 Old leaf of young turnip plant. Note backward curling and bright red and yellow margin. (Peat culture)

34 Corky stem on 12 week old broccoli plant. (Peat culture) See footnote.

35 Typical cracked, corky area on upper midrib of cabbage. (Peat culture)

36 Rosetting of cabbage. (Peat culture)

37 Close-up of same plant showing hooking and marginal scorch of young leaves. Note broad, swollen midribs.

29

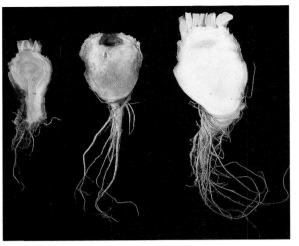

38

41

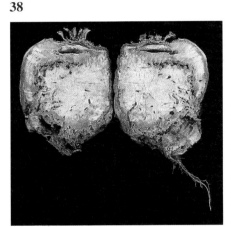

39

42

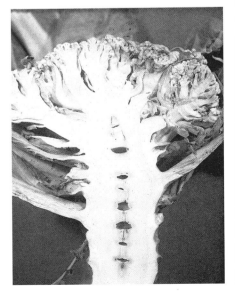

40

43

(Boron deficiency, cont'd)

38 Left and centre: swede roots showing brown heart cut through vertically. Note discoloration and hollow below crown of centre root. Right: healthy root. (Peat culture)

39 Turnip root cut open to show hollows, fibrous condition and advanced rotting following brown heart. (Peat culture)

40 Hollow stem of cauliflower.

41 External appearance of root of 15 week old swede plant showing leathery appearance. Note also corkiness on petioles, one of which has been cut off to expose the others. The deformed growing point (arrowed) is also visible. (Peat culture)

42 Brown curd of cauliflower.

Note: Similar cracking and corkiness on stems and petioles can result from the action of growth regulator type herbicides (see p. 12).

Molybdenum deficiency

Of the brassica crops, cauliflower (both summer and winter types) is the most sensitive to molybdenum deficiency. Other species are not normally affected, and the following description is based on cauliflower.

Young plants, interveinal chlorosis or whitening, especially near leaf margins; leaves cupped and elongated. In older plants, 'whiptail', a serious disorder of commercial crops in which lamina of younger leaves is progressively distorted, brittle and greatly reduced, eventually leaving largely bare midrib. Growing point becomes blind or stub-like and new shoots may appear from hypocotyl. Although molybdenum deficiency normally occurs on acid soils (below pH 5.0), whiptail of cauliflower can occur up to pH 7.0. It is associated with leaf Mo concentrations of less than 0.13 μg g^{-1} in dry matter (Plant, 1956).

43 Young plants with chlorosis and cupping.

44

45

46

47

44 'Whiptail'. Youngest (most affected) leaves arrowed.

45 Plant showing signs of recovery (new growth in centre) probably as a result of transplanting into non-deficient soil (Mo concentration in leaves 4.3 µg g^{-1} i.e. no longer deficient). (Field crop)

46 'Blindness' due to cold shock (see footnote). Such plants do not recover. (Seedling raised in field soil in polythene tunnel)

Note: Blindness of the growing point, with some swelling of the surrounding petiole bases, can also be caused by preceding sub-zero temperatures, or by swede midge attack. In the latter case very small white larvae should be present on the petioles. Both disorders are distinct from whiptail in that leaf shape is unaffected. For a fuller account of whiptail occurrence in the field see Plant (1951, 1956).

Manganese toxicity *(soil acidity complex)*

Concave cupping of leaves (particularly medium-old) and pale 'rim' round leaf edges, followed by marginal spotting and scorching. Feint interveinal mottling. In field, where toxicity is a result of soil acidity, additional symptoms associated with calcium deficiency, such as convex cupping and tip hooking of young leaves. Mn concentration in leaf margins exceeds 100 µg g^{-1}.

47 Swede showing cupping and light rim on leaves. (Sand culture)

Note: For further details see Hewitt (1945, 1946) and volume 1, p. 97.

Broad Bean

48

49

50

Nitrogen deficiency

This crop nodulates very readily and nitrogen deficiency does not therefore occur in normal practice. *(not illustrated)*

Phosphorus deficiency

Stems thin, leaves erect; early defoliation of basal leaves.

48 Two plants showing all these symptoms. (Sand culture)

Sulphur deficiency

All leaves pale yellowish-green. Leaflets become increasingly erect, giving 'Christmas tree' effect.

49 14 week old plants showing both symptoms. (Sand culture)

Note: P deficiency can also result in erect leaves.

Potassium deficiency

Speckling of older leaf margins leading to marginal scorching, wilting and abscission. Scorched edges may curl upwards. Internodes may be short.

50 Complete young leaf with slight speckling and an old leaflet with severe scorching. (Sand culture)

51
52

Calcium deficiency

Death and blackening of growing point; or collapse of stems and petioles; or pods deformed, wilted and blackened, seeds fail to develop.

51 Calcium withheld from start in NVRS sand culture experiment: shoots emerge blackened. (This was also observed by Hewitt, 1945). Healthy growth recommenced after restoration of calcium supply.

52 13 week old plant: collapse of stems and petioles. Mature leaves contained 0.021% Ca. Healthy 0.13% Ca. These very low values were confirmed by repeat analysis. Leaves of field-grown beans normally contain 1–2% Ca. (Sand culture)

53 Symptoms on pods and seeds.

Note: Boron deficiency can also cause stem collapse.

53

Magnesium deficiency

In young plants, central interveinal chlorosis, margins green: later on, very pronounced interveinal chlorosis, worst at margins.

54 Third and fourth leaf pairs of five week old plant showing central interveinal chlorosis: mean Mg% = 0.039; healthy 0.20%. (Sand culture)

55 Nine week old plants, showing marginal form of chlorosis. (Sand culture)

54

55

56

Iron deficiency

New leaves pale green or
yellowing, with necrotic areas
predominantly around margins of
older leaves with occasional
interveinal necrosis near midrib.

56 Plant showing pale young
leaves and necrosis. (Flowing
solution culture)

Manganese deficiency

57 Brown lesions in centres of
cotyledons, similar to 'Marsh
Spot' in peas: leaves practically
normal or slight interveinal
chlorosis.

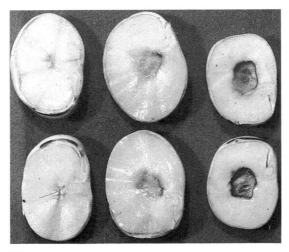

57

Zinc deficiency

Young leaves erect, tips curled
back, pointed and rolled inwards
with wavy margins. Subsequent
leaf expansion severely restricted.
Flowers dehisce prematurely
without setting pods. Growing
point dies.

58 Plant grown in flowing solution
culture, showing leaf and flower
symptoms.

Copper deficiency

In sand culture, Hewitt (1951)
observed erect, spindly habit;
leaflets narrow, pointed and
curled up at tip: wilting and
withering of leaf tips followed.
Red-brown necrotic areas
developed. For field incidence,
MAFF (1976) says young leaves
grey-green or chlorotic. Flowers
lose dark purple-brown pigment
and become pale brown.

58

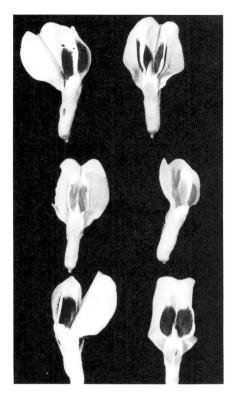

59

60

59 Central pair of flowers grown from copper-deficient seed in absence of copper: upper pair from normal seed in absence of copper: lower pair from normal seed in complete nutrient. Note reduction of pigment in first case.

Boron deficiency

Interveinal chlorosis of medium-old leaves starting at edge and moving inwards. Young leaves cupped and deformed: stem may collapse, as shown for calcium deficiency in Plate **52**.
Growing points may die with consequent outgrowth of laterals from base. Brenchley and Warington (1927) mention roots short and stumpy; failure of nodulation; dead flower buds.

60 Young leaves of eight week old plants, showing cupping. (Peat culture)

61 Same plant showing chlorosis on lower leaves.

61

Carrot

62

63

65

64

66

Nitrogen deficiency

Foliage uniformly pale green and has frail appearance due to fineness of leaflets. Oldest leaves become yellow, sometimes with red tints, and shrivel.

62 Field appearance of mildly deficient crop. Note fine, pale foliage. (N V R s field experiment with N levels).

63 Right, typical deficient plant and left, healthy plant from same experiment. Note relatively small difference in root size.

64 Right, deficient mature leaf and left, corresponding healthy leaf from same experiment to illustrate difference in colour, fineness of leaflets and internode length. The deficient leaf petiole tested 0 for nitrate: healthy leaf $100 \ \mu g \ ml^{-1} \ NO_3$. (All pictures taken mid-September).

Phosphorus deficiency

Purpling of older leaves in young crop, beginning at margin: also purpling of older petioles and stunting. No yellowing.

65 Nine week old plants, showing leaf purpling. (Sand culture)

Note: Possibility of confusion with carrot fly attack: examine taproots for signs of gnawing. Carrot motley dwarf virus causes older leaves to redden or purple but in this case younger leaves are yellow.

Sulphur deficiency

New leaves uniform yellow colour, and may appear rather frail.

66 Nine week old plant, showing pale new leaves. S% in young leaves of ten week old plant was 0.19; healthy 0.28. (Sand culture)

Note: Resembles Mn deficiency.

67

69

70

68

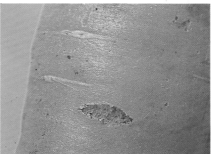

71

Potassium deficiency

Old leaves scorch and collapse, beginning at margins of leaflets. Later, entire petioles of these leaves acquire water-soaked appearance, then dry up and collapse.

67 Nine week old plant showing complete collapse and death of old leaves. Mean K% of second leaf was 0.32; healthy 1.9%. (Sand culture)

Calcium deficiency

Necrosis of growing point and new leaves. Sudden appearance of short lengths of watersoaked area on petioles, leading to collapse of distal part whilst still green. Later, these leaves shrivel up. In NVRS sand culture experiment, roots had brown core (Plate **70**).

68 Growing point necrosis on four week old plant. (Sand culture)

69 Petiole collapse in thirteen week old plant. (Sand culture) Calcium supply was stopped four weeks previously. Ca% in fourth pair of leaves was 0.13.

70 Brown core in roots of thirteen week old plants. (Sand culture)

71 'Cavity spot'. According to Maynard et al. (1961) this is due to calcium deficiency, but it was not observed in the NVRS experiment, nor in UK work by Needham (1971). (Field crop)

Note: By contrast with potassium deficiency, the watersoaking of petioles is confined to a short length, and distal part, including leaf, stays green initially.

72

Magnesium deficiency

Older leaves chlorotic beginning at edges. Red tints move in from margins. Some backward curling of leaflets may occur.

72 Left to right: recently mature, intermediate and old leaves of ten week old plant showing marginal chlorosis (centre) and reddening (right). Mg percentages 0.058, 0.040 and 0.038 respectively; healthy, old leaf 0.35%. (Sand culture)

Notes: Easily confused with nitrogen deficiency and carrot motley dwarf disease. Nitrogen deficient plants have a rather uniformly pale appearance, with fine leaflets and give a low reading for petiole nitrate. In magnesium deficient plants the new growth is much greener than the old, and the foliage does not have a particularly fine appearance. Such plants should have a high nitrate status until mid-season. Motley dwarf disease, in Britain, is caused by two viruses, carrot mottle virus and red leaf virus (Watson et al., 1964). These produce a range of colours from purple, through salmon pink to yellow in both old and new leaves, depending on genotype. Autumn King types show virtually no symptoms. Virus attack may begin in the very young crop: this rarely happens with magnesium deficiency. If plants subjected to reduced competition e.g. those on the edge of beds, are healthier than strongly competing plants, a nutritional explanation is likely.

73

Manganese deficiency

Whole plant, but especially young leaves, uniformly pale yellowish-green and frail. Sap nitrate high. Appearance in field is a patchy distribution of bright yellow areas whilst the rest of the field appears a reasonably normal green.

73 Mature leaves from healthy (left) and deficient plants. Mn concentrations 65 $\mu g\, g^{-1}$ and 0 $\mu g\, g^{-1}$ respectively. (Peat culture)

74

75

Copper deficiency

Youngest leaves become very dark green and fail to unfold. Older leaves appear wilted. Most serious on peat, regardless of pH.

74 Copper deficient plant grown in solution culture, illustrating non-opening of leaves.

75 Right: crop failure on peat soil at Lullymore, Co Kildare, Eire. (No copper given.) Left: copper sulphate applied as a side dressing at 10 kg ha^{-1} Cu.

Note: On English fen peat soils, carrots have responded to copper treatment (including residues of 10 kg ha^{-1} Cu applied ten years previously although visual symptoms of deficiency were not present. (Pizer et al., 1966.)

Boron deficiency

Leaflets of young leaves greatly reduced (Plate **78**, right) and later die back (Plate **78**, left).
Old leaves chlorotic, curled backwards, giving prostrate habit, with purple edges to leaflets (Plate **79**, right). Growing point may die. Corky splits may occur on petioles.
    Roots split, exposing central core: surface of root dull, greyish. Core may contain hollows.
    A disease known as '5 o'clock shadow' is at least partially due to boron deficiency (MAFF, 1976). After steam cleaning, the roots are seen to be covered in numerous small black spots giving a blotchy appearance. This greatly affects their value, especially for canning.

76 Split roots of 11 week old plants: note that these roots were not cut: the cortex has split leaving 180° opening. (Peat culture)
77 Left, intact surface of B-deficient root (split at back) showing dull appearance. Right: healthy root. (Peat culture)
78 Right, young leaf; left, slightly older necrotic leaf of similar age. Centre, healthy leaf. (Peat culture)

76

77   **78**

79

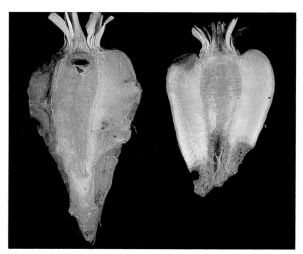

80

81

(Boron deficiency, cont'd).

79 Old leaves of 11 week old plant. Right: B-deficient. Left: healthy leaves. (Peat culture)

80 Hollows in core and breakdown of cortex. (Peat culture)

81 'Five o'clock shadow'. Note discoloured areas. (Field crop)

Note: Boron application to the soil may not prevent splitting in the field: in a field trial at N V R S in 1980, both 'Solubor' (11 kg ha^{-1}) and basal N applications greater than 40 kg ha^{-1} significantly increased splitting.

Celery

82

Healthy plant

82 13 week old plant. (Sand culture)

Nitrogen deficiency

Older leaves turn uniform yellow and eventually white.

83 Field crop of cv 'Jason' growing on peat soil. Note pale crop colour on bed in centre of picture due to N deficiency caused by excessive irrigation on this bed. Marketable yield in this bed was severely reduced.

84 Close-up of pale leaves in the bed mentioned above. Petioles of such leaves gave nitrate sap test readings of 0 compared with about 100 μg ml^{-1} NO$_3$ for corresponding petioles in greener beds.

85 Left to right: young, medium and old leaves of same plant showing progressive chlorosis, which is always uniform on a particular leaf. Apparently darker areas on right hand leaf are shadows.

83

84

85

86

87

88

Phosphorus deficiency

Growth dwarfed and spindly: foliage dull bluish-green, older leaves yellow and die early.

86 13 week old plant showing yellowing old leaves. Two mature leaves contained 0.36% P; healthy 1.5% P. (Sand culture)

Note: Very similar to N deficiency. Distinguish using nitrate test strips: if P deficient, petiole sap should give high reading for nitrate.

Sulphur deficiency

Uniform yellowish-green leaves, especially on new growth.

87 13 week old plant. Note paler new leaves in middle. Two mature leaves contained 0.09% S; healthy 0.27% S. (Sand culture)

Note: Young leaves paler than old (contrast with N, P and Mg deficiency). Nitrate sap test should give high reading.

Potassium deficiency

Growth stunted; leaves convex, very shiny, dark green; slight marginal and interveinal paling with pale brown interveinal necrotic spotting (Hewitt, 1945).

88 Terminal leaflets showing necrotic spotting.

Note: Mature leaves of three month old plant showing deficiency in NVRS experiment contained 0.73% K; healthy 3.0% K.

Calcium deficiency

Field symptom is 'blackheart'—blackening and death of growing point. The disorder is normally induced by an excess of other cations, or drought. When the excess cation is potassium, growth is bushy, and leaves bluish-green: if sodium, plants are tall, and leaves medium green. In young plants in sand culture, leaves collapsed following death of tissue near junction with petiole.

89 Symptoms seen in six week old plant. Left: death of leaflet tissue near junction with petiole; centre: death of growing point; right: complete death of leaf and necrosis moving down petiole. (Sand culture)

90 'Blackheart' in 13 week old plant, from which calcium has been withheld for one month. Two mature leaves contained 0.22% Ca; healthy 0.92% Ca. (Sand culture)

Note: See Geraldson (1954) for further information. See also remarks about boron deficiency on peat soils in Ireland (Appendix 2).

89

90

91

Magnesium deficiency

Chlorosis of older leaves beginning at margin and moving in until, on oldest leaves, entire leaf is white. At this stage margins may show feint pink tint.

91 Sequence of leaflets taken from young green leaves (top left) to old, white leaves (bottom right), showing each stage in the development of the chlorosis.

Note: Distinguished from N deficiency by the striking marginal chlorosis on leaves in intermediate positions. With N deficiency, these are uniformly yellow-green. For further information see Yamaguchi, Takatori and Lorenz (1960).

Manganese deficiency

92 Interveinal chlorosis, worst on old leaves. Left, healthy leaf. Centre and right, deficient leaves.

92

93

Boron deficiency

Celery is prone to boron deficiency and numerous symptoms occur, not necessarily simultaneously. 'Cracked stem' (Purvis and Ruprecht, 1937) is when small transverse cracks appear on the outer surface of the petiole, the epidermis next to the crack curls back and the damaged area goes brown. In severe cases this can give a shredded appearance to the plant. 'Brown checking' (Yamaguchi and Minges, 1956) is the development of brown, watersoaked oblong spots, with or without transverse cracks on the inner petiole surface. These may extend up to the full length of the petiole. Other symptoms (Hewitt, 1945) include development of axillary shoots, with or without death of the growing point, distortion of young leaves, browning of leaf margins and twisting of petioles. Varietal differences in susceptibility occur.

93 Young leaf symptoms: centre leaves glossy dark green, curled downward; older leaves pale brown margins. (Peat culture)

94 Field grown plant with several symptoms. This picture shows axillary shoots and a large crack low down on a main petiole (centre). The growing point was still intact.

95 Beginning of stem cracking on outer petiole of same plant. Such cracks may become very numerous and turn brown.

94

95

96

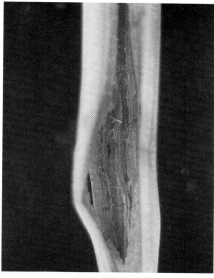

97

96 Curled, wrinkled secondary petiole (same plant).
97 Brown lesions on inner petiole surface of same plant.

Note: See also appendix 2.

Leek

The response of leeks to all kinds of disorders is rather similar, and the differences have proved very difficult to illustrate photographically. Leaf analysis is likely to be necessary in many cases for a reliable diagnosis. The following notes are based largely on observations made on the NVRS experiments.

Nitrogen deficiency

Old leaves fade to uniformly pale green near the tip, which eventually shrivels. Leaves erect and narrow: growth severely reduced. Severely deficient 13 week old plants contained 1.0% N; healthy plants 3.4% N (Whole plants).

Phosphorus deficiency

Growth dwarfed and thin, leaves dull blue-green; tips die back.

Sulphur deficiency

Leaves very stiff and erect. Early swelling of stem base. With extreme deficiency, the new leaves are short, thickened and chlorotic. Deficient 13 week old plants contained 0.17% S; healthy plants 0.40% S (Whole plants).

Potassium deficiency

Dieback of old leaf tips.

Calcium deficiency

Leaves very narrow and die back abruptly from tip without first yellowing.

Magnesium deficiency

Older leaves yellowish-green, particularly near base. Deficient plants contained 0.043% Mg; healthy plants 0.18% Mg (Whole plants).

Boron deficiency

Transverse cracks on leaves.

Lettuce

98

99

100

101

Healthy plant

98 Mature field-grown plant (cv 'Great Lakes').

Nitrogen deficiency

Pale colour, very reduced growth, small hearts. In extreme cases, outer leaves turn pale yellow and die.

99 Deficient plant of same age, variety etc. as in Plate **98**. Note smaller size.

100 18 day old plants showing large reduction in growth despite only slight yellowing of cotyledons. Left, healthy plant: right, N-Deficient plant. Sap nitrate concentrations, 4000 μg ml^{-1} (healthy plant), 0 μg ml^{-1} (deficient plant) (midrib of large leaf): Total N (whole plant) was 4.7% in healthy plants and 2.0% in N-deficient ones. (Sand culture)

101 Severely deficient 6 week old plant cv 'Lobjoits Cos' grown in Perlite/peat mixture. Petiole sap nitrate concentration 5 μg ml^{-1}.

Note: The colour difference due to cultivar is much greater than that due to N effects so that the colour symptom is only really useful for diagnosis if the comparison is made of old and young leaves on the same crop.

Phosphorus deficiency

Growth is very severely reduced and plants fail to heart but other symptoms practically non-existent. Old leaves may present a somewhat dull or bronzed appearance. *(Not illustrated)*

Note: Leaf analysis is the most reliable guide to P deficiency.

102

103

Sulphur deficiency

Leaves, especially young ones, small, stiff and pale, yellowish-green, giving rosetted appearance and causing severe stunting.

102 Six week old plant. (Sand culture)

Potassium deficiency

Older leaves severe marginal and interveinal scorch.

103 Old leaf from cv 'Great Lakes' showing neat pale brown scorch along entire leaf margin. K concentration 0.38%. (Sand culture)

104

105

Calcium deficiency

Puckering and necrosis of margins of young leaves and sometimes death of growing point.

104 Six week old plant, showing severe leaf distortion and necrosis. (Sand culture)

105 Nine week old plant, showing tip-burn-like symptoms, see footnote.

Note: 'Tip-burn' of lettuce is very similar to these symptoms and has been induced in sand-culture by withholding calcium (Ashkar and Ries, 1971).

In commercial practice, however, it is associated with many other factors. See p. 10.

106

107

Magnesium deficiency

Older leaves chlorotic marbling, especially on distal part.

106 Second, third and fourth leaf of four week old plant, showing early symptoms. Mean Mg = 0.038%; healthy 0.33%. (Sand culture)

107 Field grown lettuce originally diagnosed as Mg deficient but which could equally well be virus infected (see footnote overleaf).

108

109

(*Magnesium deficiency, cont'd*)

108 Lower leaf of cv 'Cobham Green' inoculated with beet western yellows virus. See footnote. (Field grown).

Note: Strikingly similar to beet western yellows virus—see Tomlinson and Webb (1978). The lowest leaves of virus-infected plants have a hard, parchment-like feel, and a neat brown marginal scorch (Plate **108**), whereas leaves of magnesium deficient plants are pliable and not normally scorched.

Manganese deficiency

Whole plant, especially older leaves, pale. Later, interveinal grey-brown necrotic spotting on older leaves.

109 Deficient plant growing in field.

110

Zinc deficiency

When grown in solution culture, deficient plants are small and rosetted, with large papery necrotic areas, with a dark margin, between veins (Van Eysinga and Smilde, 1971). Old leaves first affected. In California growth responses to zinc were obtained in the field although no symptoms were seen (Zink, 1966).

110 Symptoms on cv Deci-Minor in solution culture.

Copper deficiency

Leaves elongated, chlorotic at edges. Young leaves cup, but chlorotic edges curl downwards. Later, leaves wilt, beginning at edge and tip. Growth severely reduced: no heads formed. (M A F F, 1976, Van Eysinga and Smilde, 1971.)

111 Note elongated leaves and marginal chlorosis.

111

112

Boron deficiency

Tipburn-necrosis and puckering of leaf edges very similar to calcium deficiency (Plate **105**). With boron deficiency, the necrosis becomes progressively worse as one approaches the growing point, which is completely blackened, causing failure of hearting and hence rosetting: also the young leaves are more deformed and brittle with prominent midribs.

112 Growing point death and leaf margin puckering in nine week old plant. (Peat culture)

Molybdenum deficiency

Van Eysinga and Smilde (1971) report that young plants turn pale green: leaf margins turn brownish-yellow and wither. Older plants show translucent spots which become necrotic and coalesce. Oldest leaves affected first. Growth very stunted and plants may collapse entirely.

113 Two leaves of cv Deciso grown in peat block, showing translucent patches.

Note: Observed in Holland on acid, ironstone-containing soils with 60% organic matter (Mulder, 1954), and in Britain on a mineral soil of pH 5.2 (Plant, 1956).

113

Manganese toxicity

Irregular pale yellow margins to older leaves, sharply demarcated from rest of leaf which stays green.

114 Mn toxicity observed on a sandy loam soil, pH 5.0, at NVRS in 1972. Mn concentration in outer leaves, $650\,\mu g\,g^{-1}$. Unaffected plants where soil pH was 6.9 contained $50\,\mu g\,g^{-1}$.

114

115

116

117

Boron toxicity

In young plants, neat pale yellow margin around edge of leaves with no other symptom; or (large plants, Van Eysinga and Smilde, 1971) brownish-grey sunken spots which develop into a ring-shaped pattern, with dark brown veins. The spots later coalesce and desiccate, making the leaf papery.

115 Toxicity in seedling of cv Darka on sandy loam soil, induced by application of 100 kg ha^{-1} borax. Note pale margins on older leaves.

116 Boron toxicity induced in sand culture, showing large expanses of tissue collapse on older leaves.

Chlorine toxicity

117 Marginal scorch of wrapper leaves.

Marrow

118

Healthy plant

118 Seven week old plant. (Sand culture)

Nitrogen deficiency.

Uniform yellowing of older leaves.
119 Two week old plants: left to right: healthy, N-deficient and S-deficient plants, each with cotyledon detached and tested with nitrate test strip to show high levels in healthy and S-deficient plants (dark blue test strips) and low level in N-deficient plant (pale, almost white test strip). Total N: healthy 5.3%, N-deficient 2.7% for whole plants. (Sand culture)

120 11 week old plant: note older leaves small and yellowish-green: new leaves dark green. (Sand culture)

Phosphorus deficiency

Young leaves dull emerald-green, very flat and expand very slowly.
121 Three week old plant. (Sand culture)

119

120

121

122

123

Sulphur deficiency

Young leaves uniform pale yellowish-green. See Plate **119** (Nitrogen deficiency) and caption. Total S: healthy 0.31%, S-deficient plant 0.17% for whole plants. (Sand culture)

Potassium deficiency

Uniform chlorosis and marginal scorch on old leaves which begins as feint marginal chlorosis on green leaves. Severe stunting.

122 Four week old plant showing marginal scorch on cotyledon and pale margin of true leaf. Growth ceased soon after this. (Sand culture)

123 11 week old plant. The yellow leaf contained 2.4% K when sampled ten days after the photograph was taken; healthy 5.8% K. (Sand culture)

Calcium deficiency

Tipburn and concave cupping of very young leaves, which later on cup convexly to produce a claw shape. Patchy chlorosis between veins.

124 11 day old plant showing tipburn on young leaf arrowed. (Sand culture)

125 Four week old plant, showing 'claw-shaped' young leaves with yellowing between main veins. (Sand culture)

124

125

126

127

Magnesium deficiency

Chlorotic marbling and white speckling of older leaves, leading to very severe interveinal scorch and stunting.

126 Early signs of chlorotic marbling on 18 day old plant. (Sand culture)

127 Four week old plant. Top left, first: top right, second: bottom left, third: and bottom right, fourth leaves. Mg concentrations in third and fourth leaves, 0.055 and 0.065% respectively; healthy 0.46 and 0.57%. (Sand culture)

Note: Slight resemblance to cucumber mosaic virus, although latter is on young leaves and involves distortion. (See footnote on boron deficiency, p. 54.)

Manganese deficiency

Older leaves show interveinal chlorosis leading to marginal scorch. Main veins remain green in contrast.

128 Older leaves showing marginal scorch. (Flowing solution culture)

128

129

130

131

Boron deficiency

New leaves become small, stiff, brittle and misshapen, with long jagged lobes and chlorosis between them. Young petioles crack across top at regular intervals and eventually become s-shaped, when seen from the side. Fruits crack.

129 Upper leaves of five week old plant showing brittle appearance and chlorosis. Veins prominent. (Peat culture)

130 Growing point of seven week old plant showing malformed new leaves and cracks across petioles. (Peat culture)

131 Cracked fruit. (Peat culture.)

132 Young marrow leaves infected with cucumber mosaic virus, see note. (Field crop).

Note: Mottling, distortion and crinkling of young leaves is often a consequence of infection by cucumber mosaic virus (Plate **132**), see Gibbs and Harrison (1970). The petiole cracks are highly diagnostic for B deficiency.

132

Onion

133

134

135

136

Healthy plant

133 17 week old plants (field)

Nitrogen deficiency

Growth stunted and thin; foliage pale; older leaves turn yellow and die back from tips.

134 17 week old plants (field). Spacing as in 133; bulbs actually much smaller.

Phosphorus deficiency

Poor growth. Leaves dull green: older ones die back from tip. *(Not illustrated)*

Sulphur deficiency

Leaves thick and deformed: new leaves yellow. In N V R S experiment, relatively good bulb formation despite very few leaves. *(Not illustrated)*

Potassium deficiency

Older leaves die back from tip without first becoming yellow. Loss of turgidity.

135 13 week old plants. Total K in mature leaves 0.80%; healthy 4.1%. (Sand culture)

Calcium deficiency

Die back of young leaves without prior yellowing, or death of short length of leaf causing distal part to topple over and die.

136 13 week old plant, with new leaf emerging dead. Total Ca in mature leaves 0.18%; healthy 0.52%. (Sand culture)

137

138

Magnesium deficiency

Older leaves turn uniform yellow along entire length, without die back.

137 17 week old plant. Total Mg in whole plant 0.034%; healthy 0.37%. (Sand culture)

Note: Use of nitrate test strips helps to distinguish this from N deficiency.

Manganese deficiency

Striped chlorosis of outer leaves, followed by necrosis. Growth severely reduced. *(Not illustrated)*

Note: See Zn deficiency.

Zinc deficiency

Leaves striped yellow; twisted; stunted. Fairly common in USA. *(Not illustrated)*

Note: See Mn deficiency.

Copper deficiency

Tips of youngest leaves become chlorotic, turn white and twist into spiral or bend at right angles to rest of plant. Bulb scales pale yellow and thin. Bulbs lack solidity. *(Not illustrated)*

Boron deficiency

Older leaves become chlorotic and die back. Regularly spaced feint yellow lines across leaves, which develop into cracks. One or two isolated cracks 3–4 cm from leaf tips may appear earlier than this. Stuart and Griffin (1944) also mention deep blue-green leaves, and youngest leaves conspicuously mottled yellow with distorted shrunken areas.

138 Leaf cracking (arrowed) in 17 week old plant. (Peat culture)

Molybdenum deficiency

Poor emergence and death of seedlings on peat soils of pH below 5.4 (see p. 88). In fully grown plants the leaf tips die and there is a wilted flabby zone between the dead tip and the healthy green part of the leaf. *(Not illustrated)*

Parsnip

139

140

141

142

Healthy plant

139 Six week old plant. (Sand culture)

Nitrogen deficiency

Leaves small, pale green, and lustreless; weak spindly growth; small roots.

140 Ten day old plants showing big effect of N on size despite lack of symptoms. Left: healthy plant (petiole sap nitrate 7000 µg ml^{-1}). Right: N-deficient plant (0 µg ml^{-1}), both figures for large leaves. Total N in whole plants: healthy 3.1%, N-deficient plant plant 2.7%. (Sand culture)

141 Seven week old plants. Left: healthy plant (petiole sap nitrate 4000 µg ml^{-1}). Right: N-deficient plant (15 µg ml^{-1}), both figures for lowest leaves. (Sand culture)

Phosphorus deficiency

Spindly growth; thin petioles; dull leaves. Hewitt (1944) observed purple tints on leaves and petioles, small thin roots.

142 13 week old plant. Note slender petioles. Two mature leaves contained 0.13% P; healthy 1.1% P. (Sand culture)

143

Sulphur deficiency

New leaves uniformly pale yellowish-green, stiff, and slightly concave. Such leaves appear to have sharply toothed margins, and a fine network of recessed veins.

143 Nine week old plant showing young leaf (left) much lighter green than old ones. (Sand culture)

Potassium deficiency

Marginal and interveinal chlorosis of older leaves leading to scorch: margins of scorched leaflets roll upward.

144 Young leaf (left) and older (third) leaf (right) of ten week old plant. Latter shows chlorotic margins, and contained 1.6% K; corresponding leaf from healthy plant 5.6% K. (Sand culture)

Note: The chlorosis symptom illustrated here is extremely like one form of magnesium deficiency. Leaf analysis would be required to distinguish.

144

145

Calcium deficiency

Leaflets show collapse of tissues near junction with petiole or petiole itself becomes watersoaked. Affected area then dries out and so remainder of leaf or leaflet hangs down and eventually dies. Growing point may die.

145 Five week old plant dismembered to show different stages of the disorder. Top left: necrotic growing point. Top right: youngest leaf showing early tissue collapse with moist fringe separating dying tissue from healthy surrounding leaf. Bottom left: next youngest leaf— collapsed area has dried out. Bottom right: oldest leaf and most of petiole shrivelled up. (Sand culture)

Magnesium deficiency

Severe interveinal chlorosis of older leaves, sometimes beginning at margins and sometimes near middle. Later, badly chlorotic areas scorch.

146 Left to right: fourth, third and second leaves of 13 week old

146

plant with chlorosis and scorched areas towards centre of leaf. Mg percentages, respectively, 0.080, 0.054, 0.065; healthy plant (leaf 3) contained 0.35% Mg. (Sand culture)

Note: The marginal form of chlorosis shown for K deficiency (Plate 144) also frequently results from Mg Deficiency. Several parsnip viruses can also induce similar chloroses.

Manganese deficiency

Marginal and interveinal chlorosis of nearly all leaves: chlorotic areas olive green.

147 Typical affected leaf.

Note: Differs from Mg and K deficiency in that the chlorotic areas are light green rather than yellow and in the whole plant being fairly uniformly affected.

147   **148**

149

Boron deficiency

Lateral cracks across petioles, which may break or bend as a result. New leaves glossy, old ones pale, sometimes with red margin. Roots small, discoloured inside. Growing point may die. Hewitt (1945) also observed cavity in root just below crown, and twisting and crinkling of petioles.

148 Cracked petiole on ten week old plant. (Peat culture)

149 Chlorosis on old leaf of same plant. Note feint red margin. (Peat culture)
150 Roots cut through: left, healthy; right, B-deficient. In the latter, note the discoloration around the central xylem tissues and the lateral roots arising from them (arrowed). (Peat culture)

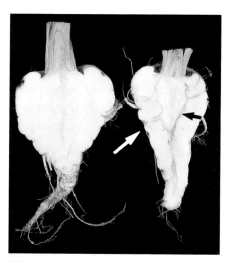

150

Pea

151

153

152

154

Nitrogen deficiency

Peas normally nodulate readily and therefore do not show N deficiency symptoms. Before this occurs, deficient plants appear frail and are a rather light green.

151 Three week old plants: left, deficient; right, healthy. (Sand culture)

Phosphorus deficiency

Slow growth; small leaves, thin stem; upper leaflets erect: lower leaves shrivel, beginning at edge. Hewitt (1944) also observed small, dull, pale olive leaves, and few, poorly filled pods.

152 Four week old plant showing erect leaves (A) and shrivelling lower leaves (B). This plant later shrivelled completely. (Sand culture)

Sulphur deficiency

New leaves very small: plant small and frail, with 'vertical' appearance because petioles slope up at 45°. Foliage yellowish-green.

153 Left with, and right without S.

Potassium deficiency

Young internodes very short and plant squat; older leaves marginal scorch.

154 Five week old plant showing foreshortening of internodes near top of plant. (Sand culture)

155

156

157

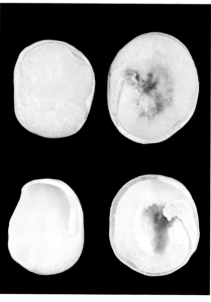

158

159

Calcium deficiency

Young stems, pedicels and leaf tissues wilt and collapse; interveinal chlorosis; pods and seeds imperfectly developed.

155 Note collapse of top of plant and general desiccation.

Magnesium deficiency

Interveinal chlorosis on older leaves, beginning near centre of leaf-let and so giving a characteristic green margin which distinguishes it from manganese deficiency.

156 Typical chlorotic leaf in a mature field crop. In NVRS experiment chlorotic leaves contained 0.042% Mg, compared with 0.24% Mg in healthy plants.

Iron deficiency

New leaves uniform pale yellow, with irregular necrotic spotting, tendrils chlorotic, older leaves remaining green (Seeliger and Moss, 1976)

157 Iron chlorosis induced in solution culture.

Note: Resembles 'top yellows' — a disease caused by bean leafroll virus.

Manganese deficiency

Haulm may appear normal or leaves show slight interveinal chlorosis: cut surfaces of seeds have brown spot or cavity in centre (Marsh spot) particularly when dried and soaked. In mild cases the spot is simply granular tissue, without discoloration.

158 Leaves of field crop, showing interveinal chlorosis beginning at margins.

159 Two forms of marsh spot (right): healthy peas on left. (Peat culture)

Note: Common and economically important in English pea growing areas when pH exceeds 6.1 on soils with high organic matter e.g. old permanent pasture, peat, or marshland soils. May also occur where waterlogging results in poor root system development followed by drought which induces the disorder. (MAFF, 1976.) 61

160

Zinc deficiency

Lower leaves necrotic margins, stems stiff and erect: flowers absent (Piper, 1940b).
(Not illustrated)

Copper deficiency

According to MAFF (1976) young leaves become greyish-green and then chlorotic. Plant wilts and withers, with severe loss of seed yield. May occur on chalky or peat soils. Hewitt (1951) found that pods enlarged normally but were devoid of seed.

160 Pods, externally normal (left) but containing no seed.

Boron deficiency

Vegetative shoots grow out from main axils: young leaflets and tendrils much reduced in size, and shrivelled. Leaves chlorotic, growth feeble. Eventually whole plant dries up and new green shoots arise at base. Pods contain very few peas, which are deformed and very variable in size.

With more extreme deficiency Wallace (1961) found stiff, thickened stems: squat bushy habit: chlorotic foliage: young leaflets small with brown tips: dead growing points. Hewitt (1945) found pods to be small and distorted, and young leaves wrinkled.

161 Nine week old plant; note axillary shoots. (Peat culture)
162 Pods: right B-deficient; left healthy. (Peat culture)

161

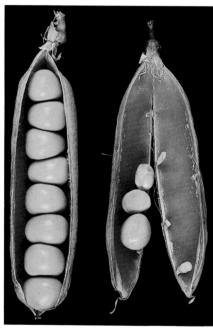

162

Phaseolus beans

163

164

165

Nitrogen deficiency

All except very young leaves uniformly pale green: oldest may turn yellow in severe deficiency. Growth very reduced; few flowers. In runner beans, midribs and veins may redden. In temperate climates these species, although leguminous, do not normally nodulate sufficiently well to enable them to give maximum yield without an external N supply.

163 Nine week old French bean plant. Note pale leaves, becoming yellow and drying up. (Sand culture.)

Note: See also Plate **164** (comparison with P deficiency).

Phosphorus deficiency

Growth dwarfed and thin: leaves dark, lustreless green: older leaves turn brown and are shed early.

164 French bean: terminal leaflets of mature leaves of, left healthy, centre N-deficient and right P-deficient plants at nine weeks of age. Note dull appearance of latter. (Sand culture)

Sulphur deficiency

In French bean, (and probably therefore in runner bean) uniform golden yellow chlorosis of new leaves. 'Suspended animation' in that symptoms do not get worse, but growth stops as newest leaves fail to expand.

165 Five week old plant showing yellow trifoliate leaf against green primary leaf below. (Sand culture)

Notes: By contrast with N deficiency, it is the new leaves which are pale. Also, petiole sap nitrate levels are high e.g. 4000 $\mu g\ ml^{-1}$ for trifoliate leaf shown here.

166

Potassium deficiency

Marginal chlorosis of old leaves followed by marginal scorch moving in between veins. Slight convex curvature to leaflet but scorched edges may curl upwards.

166 Left, first trifoliate leaf, (showing all above symptoms) and right, primary leaf of five week old French bean plant. Trifoliate leaf contained 0.36% K; healthy leaf 4.0% K. (Sand culture)

Note: Certain herbicides can cause rather similar symptoms.

Calcium deficiency

The predominant symptom is loss of turgor: hence leaves wilt and pods turn soft and yellow. (Hewitt, 1945.) In NVRS experiment, wilting of leaves began with chlorosis and tissue collapse near base of leaflet (Plate **167**). Old leaves scorch and fall off. Pods deformed, seeds fail to develop. Runner beans also have pale green leaves with chlorotic patches and brown spots at tips and margins. French bean seedlings may show hypocotyl necrosis.

167 Five week old French bean plant showing wilted leaves, chlorosis near leaf base (A) and extensive tissue collapse in corresponding area of older leaf (B). (Sand culture)

168 Leaf symptoms on runner bean.

169 Poor development of seeds (French bean).

170 Hypocotyl necrosis in French bean (arrowed).

Note: See notes on boron deficiency for hypocotyl disorders: also Shannon, Natti and Atkin (1967) for details of hypocotyl necrosis.

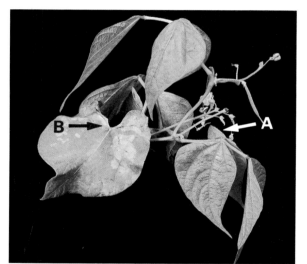

167

169

168

170

171

172

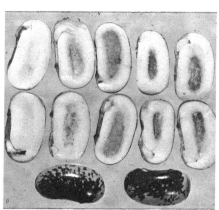

173

174

175

Magnesium deficiency

Rusty speckling (bronzing) of interveinal areas on upper surface of older leaves. Spots are about 0.5 mm across and appear angular and slightly sunken when magnified. Eventually the bronzed areas blacken.

171 Interveinal bronzing in runner bean.

172 Advanced necrosis in runner bean.

Iron deficiency

Young leaves pale green to bright yellow, showing profuse irregular necrosis, especially in acutely chlorotic leaflets. Fully expanded leaflets curve convexly and wither at tips. Oldest leaves dark green. Death of growing point and abscission of youngest unexpanded leaflets in cases of severe deficiency. (French bean).

173 French bean growing in flowing culture, showing yellowing, necrosis and curvature of young leaves.

Manganese deficiency

Interveinal chlorosis beginning as a very fine speckling on young leaves, which also appear 'pimply' from close up. (Plate **175,** left). As the leaves grow, they become smooth and the chlorosis is clearer, but less speckly. (Plate **175,** right).

Seeds show 'marsh spot' or pods may be yellow and unfilled (Hewitt, 1944).

174 'Marsh spot' in runner bean.

175 Leaves of deficient 11 week old French beans grown in peat. Left, young leaf (6 µg g^{-1} Mn). Right, intermediate leaf (0.3 µg g^{-1} Mn).

176

177

Zinc deficiency

French beans are particularly susceptible to zinc deficiency (Vitosh, Warncke and Lucas, 1973). These authors state that deficient plants become light green, and chlorosis moves in between veins from leaf tips and edges; new leaves may be deformed, dwarfed and crumpled, and have appearance of sun scald. These symptoms appear soon after emergence. Pods on terminal blossoms drop off. In flowing solution culture, older leaves developed distinct wavy margins, irregular necrotic areas and necrosis in veins.

176 45 day old plant grown in solution culture, showing wavy leaf margins (= crumpling?) and necrosis (= sunscald?).

Copper deficiency

In French bean, leaves are pale and irregular necrotic areas appear close to veins near base of leaflet: irregular marginal limpness and scorching (often on one side of leaflet) followed by withering and defoliation. Runner bean, leaves small, dull olive or yellow-green: irregular interveinal orange-yellow blotching on older leaves, followed by wilting and withering: chlorosis and die-back of shoot: flowering suppressed. (Both descriptions from Hewitt, 1951, p. 60).

177 Necrotic spotting in French bean.

Boron deficiency

Interveinal chlorosis on all leaves: very youngest leaves greatly reduced and curled. Growing point may die, but before this happens small axillary shoots appear but fail to grow. Stem swollen near nodes. Flowers may abort. Pods are deformed and may split: seeds fail to swell. Longitudinal cracks appear near base of stem. Deficiency is rare: these crops are very sensitive to excess boron.

178 Early chlorosis on French bean: later, only veins remain green. (Peat culture)

178

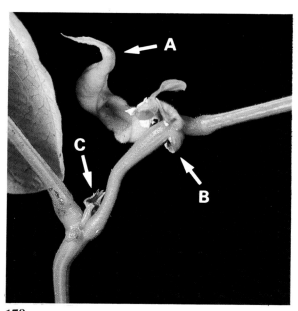

179

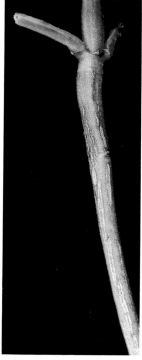

180

179 French bean plant showing: A, twisted pod split open at proximal end; B, swollen stem; and C, small axillary shoot. (Peat culture)

180 Lower part of French bean stem showing cracking and suberisation. (Peat culture)

Note: Growing point death in bean seedlings is usually due to mechanical damage to the seed (Atkin, 1958), or to attack by bean seed fly. In the latter case damage may also occur on other parts of the seedling, and careful examination should reveal larvae, which are white, and about 7 mm long, within the damaged tissue.

Manganese toxicity

The most definite and specific symptom is the presence of purple-black spots on the stem, petioles, midribs and veins, but absent from the pulvinus regions (Hewitt, 1948). These are more numerous on the ribs of the underside of the leaf than on the upper surface. Under the microscope these appear not as necrotic tissue but clumps of secreted material (actually manganese dioxide) around the basal cells of hairs (Bussler, 1958).

Other symptoms which occur on acid soils include chlorosis between major veins, most pronounced on the young leaves, which later cup convexly and scorch around the margin.

181 Black spots on petiole and veins of French bean.

182 Chlorosis and scorch. The right-hand leaf received high calcium nutrition: the other two had a lower calcium supply.

Aluminium toxicity

Chlorosis and necrosis around margins of older leaves. Common on acid tropical soils.

183 Symptoms on French bean, photographed at CIAT Colombia. (Soil culture)

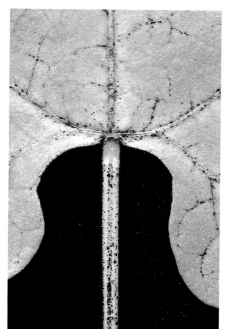

181

182

183

Radish

184

186

185

187

Healthy plant

184 14 week old plant (Sand culture).

Nitrogen deficiency

Older leaves (especially cotyledons) turn pale green and later yellow with red veins, midribs, and petioles. Growth very stunted and root fails to swell.

185 17 day old plant, with pale leaves and yellow cotyledons. Such plants contained no petiole sap nitrate as measured by test strips, and weighed about half as much as healthy plants, which contained 4000 µg ml^{-1} in large leaves. (Sand culture)

186 Acutely deficient three week old plant: old leaves yellow, with red midribs. (Sand culture)

Phosphorus deficiency

Growth dwarfed; leaves lustreless dark green, laminae and petioles dull purple tints; older leaves scorch at margins and wither early. Tendency to bolt early.

187 Plant grown in sand culture. Note scorch on leaf at centre left.

188

Sulphur deficiency

Purpling of cotyledons and old leaves, which shrivel up and absciss. Young leaves small with interveinal chlorosis. Early bolting. Root fails to swell.

188 Three week old plant showing purpling of cotyledons and small chlorotic new leaf. (Sand culture)

Potassium deficiency

Leaves small, glossy, dark bluish-green, slightly curved backward. Old leaves chlorotic in patches with margins scorched and curled upwards. Roots small.

189 Four week old plants. Left, healthy leaf, 1.8% K. Centre, mature leaf of deficient plant, 0.13% K. Right, first leaf of same plant, 0.32% K. (Sand culture)

189

Calcium deficiency

Young leaves are cupped backwards and have very neat marginal fringe of white spots which eventually turn brown. Root is normal. Wallace (1961) mentions chlorotic mottling of young leaves, tissue collapse, forward rolling of leaf margins; death of growing point. Some of these points are illustrated under Brassicas.

190 Four week old plants: Left, two leaves from deficient plant (petiole sap Ca0 μg ml^{-1}); right, healthy plant leaf (petiole sap Ca 250 μg ml^{-1}). Note convex cupping and pale marginal fringe on both deficient leaves. (Sand culture)

190

191

Magnesium deficiency

Older leaves: interveinal chlorosis, which begins as bronzing.

191 Four week old plant. Top left 1st leaf 0.032% Mg. Top right 2nd leaf 0.047% Mg. Bottom left 3rd leaf; bottom right 4th leaf. Healthy plants: 1st leaf 0.63% Mg; 2nd leaf 0.59% Mg. (Sand culture)

Notes: Could be confused with beet western yellows virus. (Duffus, 1972.) Chlorosis pattern more diffuse, and more confined to old leaves, than in manganese deficiency.

Manganese deficiency

Interveinal chlorosis of all leaves, in which the fine veins stand out clearly. Later, numerous small necrotic spots and papery areas appear. Growth, especially of root, slowed down. (Stenuit and Piot, 1960).

192 Two affected leaves: right hand one shows more advanced symptoms.

Note: See notes on magnesium deficiency.

Boron deficiency

Roots split or very thin and misshapen, with patchy, torn surface and dull skin. According to Chupp and Sherf (1960) petioles may be brittle and roots may sometimes be pale or colourless. Shorrocks (1974) says flesh of roots watersoaked with brown flecks: leaves deformed, brittle, chlorotic. Hewitt (1945) described roots as woody.

193 Left healthy, and right deficient root on six week old plants. Apart from cracks note dull, rubbery skin on latter (NVRS peat culture experiment). At this stage, no symptoms were seen on leaves.

192

193

Red Beet

194

195

196

Healthy plant

194 13 week old plant. (Sand culture)

Nitrogen deficiency

Purpling or yellowing of older leaves, associated with poor growth and small roots. Wallace (1961) mentions upright habit. In NVRS glasshouse experiment, purpling was recorded in March, and yellowing in late April— early May. In a field experiment (Plate **196**) the predominant effect was purpling. The difference is probably due to differences in temperature and light, which have a big influence on the development of pigmentation in plants. (Blank, 1947).

195 Left to right, members of third, second and first leaf pairs of 11 week old plant grown in NVRS glasshouse experiment, photographed on 25th April, showing progressive yellowing. Total N concentrations 2.4, 2.0, 1.2% respectively; corresponding leaves of healthy plant, 5.3, 4.8, 4.3%. (Sand culture)

196 Purpling of nearly all leaves on no-nitrogen plot of field experiment at NVRS, photographed on 15th September.

Note: The 'Merckoquant' test strip cannot be used satisfactorily with red beet because the sap is already purple.

197

Phosphorus deficiency

No clear diagnostic symptoms. In NVRS experiment the leaf blade appeared unusually flat, with prominent purple venation, even to the finest veins, giving a bronzed appearance. New leaves purple, small, erect.

197 Left to right: young, medium and old leaves of mature plant, showing all above symptoms. P concentration, all combined, 0.11%. (Sand culture). In field, more intense bronzing and purpling would be expected.

Sulphur deficiency

Young leaves narrow and very stiff and erect, yellowish and densely speckled with purple spots, which eventually coalesce, making whole leaf purple.

198 Ten week old plant. Note erect habit, shape and coloration of new leaves. (Sand culture)

Potassium deficiency

Old leaves become flaccid and die back from tip.

199 Five week old plant showing collapse and death of old leaves. (Sand culture)

Note: Drought can cause a similar effect.

198

199

Calcium deficiency

Young leaves have purple-black, hooked tips, which later die and exhibit leaf roll.

200 Young leaves of three week old plants showing 'hooking' and leaf roll. (Sand culture)

200

201

202

203

204

205

Magnesium deficiency

Wallace (1961) describes typical interveinal chlorosis, with reddish tinting, on old leaves. In NVRS experiment, chlorosis was not seen. Instead, we observed interveinal red mottling on leaves of intermediate age, leading to brown blisters with purple edges. In older plants in which magnesium was withheld later, there was pronounced convex 'bubbling' of the leaf lamina between the veins, so that the midrib appeared deeply recessed when observing the leaf from above.

201 Leaves 2 and 3 of five week old plant: both blistered, especially leaf 2 (left). Mg concentration, both leaves combined 0.043%; healthy 0.42%. (Sand culture)

202 13 week old plant showing 'bubbling' of leaves. (Sand culture)

Note: Ramularia disease causes purple-edged brown spots but they are smaller and more uniformly distributed.

Iron deficiency

Young leaves bleached, old leaves have red tints.

203 Mature plant showing both symptoms.

Manganese deficiency

Leaves triangular in shape, margins curled inward and severe interveinal speckling. Older leaves become chlorotic and fade to reddish colour with brown interveinal tissue. Roots reduced in size.

204 Three affected leaves from overlimed field with high organic matter (Crannymore series).

205 Manganese-deficient sugar beet leaf, to show the nature of the speckling more clearly. (Field crop)

206

207

208

Boron deficiency

'Canker' — scattered black lesions in flesh of root, sometimes with large black areas on root surface. In severe cases (but rarely in the field) new leaves are small, brittle, thick and deep red, and show 'cat-scratch' across petioles and midribs. Growing point may die causing side shoots to proliferate.

206 New growth showing petioles cracked and broken. (Peat culture)

207 Cankerous appearance of root. (Peat culture)

208 Black lesions seen in thin sections of root under water. (Peat culture)

Molybdenum deficiency

Patches of leaf die and become papery. (Wallace, 1961, p. 94).

(Not illustrated)

Spinach

209

210

211

Nitrogen deficiency

Growth stunted, lower leaves turn pale green and eventually absciss or become necrotic at tip.

209 11 day old plants showing that young plants may be very stunted by N deficiency without clear symptoms. Healthy plant (left) contained 3000 µg ml^{-1} NO$_3$ in petiole sap of upper leaves: deficient plant (right) contained none. Corresponding figures for total N in whole plant were 5.3% and 2.4%.

210 18 day old deficient plant showing yellowing and abscission of one cotyledon.

Phosphorus deficiency

Drastic reduction in growth but no other symptom. In extreme cases leaves show a slightly dull, bronzed appearance.

(Not illustrated)

Sulphur deficiency

Pale yellow-green colour, especially new leaves. Older leaves have necrotic patches near tips.

211 Three week old plant showing both symptoms. (Sand culture)

Notes:
1. As a distinction from N deficiency, S-deficient plants contain high levels of nitrate in petiole sap.
2. Could be confused with cucumber mosaic virus. (Gibbs and Harrison, 1970).

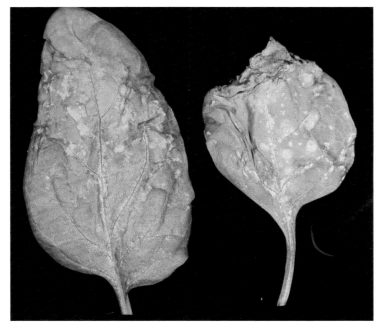

212

Potassium deficiency

Older leaves, brown necrotic patches all over, and becoming flaccid, starting at tips.

212 (Left to right). Second and third leaves of three week old plant containing, when combined, 0.22% K. Same leaves on healthy plant contained 6.7% K. Note papery necrotic patches and flaccid tips. (Sand culture)

Note: Mg deficiency produces similar necrotic patches.

Calcium deficiency

New leaves very small, distorted, twisted and chlorotic, with necrotic tips. Old leaves normal.

213 Four week old plant showing distortion of new growth. (Sand culture)

Magnesium deficiency

Patches of light brown papery tissue on older leaves. Beaumont and Snell (1935) described these as intervascular and likened them to sunscald, but in N V R S experiment the necrosis tended to follow the fine veins, as in illustration.

214 Fourth leaf of three week old plant showing large expanses of papery tissue (especially right hand side, central zone) and necrosis following fine veins. Mg concentration 0.063%. Healthy leaf 0.46%. (Sand culture)

Note: Resemblance to potassium deficiency: advisable to analyse for both elements. Hohlt and Maynard (1966) found that the 'sunscald' symptom occurred when mature leaves contained less than 0.17% Mg.

213

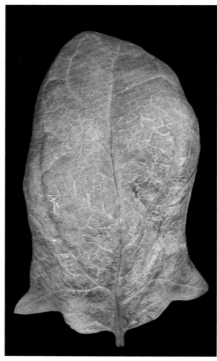

214

215

216

217

218

219

220

Iron deficiency

Laminae of young leaves yellowing progressively from bases to tips until whole leaves become golden yellow. Eventually chlorotic areas at bases and tips of leaves develop necrotic areas. Older leaves desiccate and scorch while petioles and veins remain green.

215 Plant growing in flowing solution culture exhibiting all these symptoms.

Manganese deficiency

Interveinal chlorosis of older leaves leading to crinkling and desiccation. Young leaves at growing point remain green.

216 Three affected leaves.

Zinc deficiency

Large irregular areas of sharply delineated, scorched papery tissue towards the tips of young leaves and later interveinally in older leaves.

217 Plant growing in flowing solution culture, showing scorched leaves.

Copper deficiency

Edges of young leaves become dull green, wilt and curl backwards.

218 Plant growing in flowing solution culture.

Boron deficiency

Rapidly expanding leaves are deformed—mainly cupped convexly and twisted with bulging interveinal regions. Leaf tips turn pale and eventually brown and necrotic. Older leaves become thick and very brittle. The growing point fails to develop, youngest leaves appear rudimentary and die. Side shoots grow out of axils following death of the growing point.

219 Seedling grown in peat—note stubby necrotic state of new leaves, die-back of older ones from tips.

220 Mature plant growing in flowing solution culture. Note 'bubbled' appearance of younger leaves.

Sweet Corn

221

Nitrogen deficiency

Plants are slender: leaves turn pale green with red-purple veins and edges and die back from tip. Drastic loss of yield. Leaf sheaths purple. According to Shadbolt (1959) a disorder known as 'shrivel' in which kernels, often in paired rows, fail to fill is associated with N deficiency late in crop life. It can be prevented by N top-dressing.

221 Four week old plant. Note pale, slender leaves, with tip die-back and purple sheaths. (Sand culture)

222 'Shrivel'—a severe case.

Note: See note under phosphorus deficiency.

Phosphorus deficiency

Uniform deep purpling, especially of older leaves and leaf sheaths of young plants. Seedlings very sensitive but plants may recover later. Slow growth results in delayed silking and poor pollination: hence grains arranged irregularly instead of in neat parallel rows and absent from tip of cob.

223 Lower leaves of ten week old plant. (Sand culture)

Note: In N deficiency the red shades are somewhat rusty in hue and the leaf is pale green, whereas in P deficiency the purple is more uniform and somewhat bluer. A sap nitrate test should confirm N deficiency. Poor grain fill of cobs may also result from B deficiency. Cold weather also causes purpling.

222

223

224

Sulphur deficiency

New leaves uniform golden yellow, especially towards base. Old leaf bases red.

224 Two seven week old plants showing both symptoms: Total S in third leaf 0.10%; healthy 0.89%. (Sand culture)

Note: In the case of maize, and therefore possibly sweet corn, the pattern may occasionally be interveinal (Krantz and Melsted, 1964, p. 33).

Potassium deficiency

Older leaves, marginal scorch especially at tip, eventually spreading over entire leaf: shortened internodes: small ears with pointed tips.

225 Very early signs of marginal leaf scorch.

226 Severe marginal scorch in field-grown maize.

225

Calcium deficiency

New leaves emerge twisted and shrivelled at tip and may become trapped inside blade of next oldest leaf, causing them to buckle. The tips of several leaves may remain glued together, producing a 'ladder' effect. Kawaski and Moritsugu (1979) also describe (for maize) a serration and curling of the leaf edges.

227 Four week old plant grown in sand culture. Note gelatinisation of leaf tip on right, causing buckling of next leaf to emerge. Leaf edge serration is also visible.

Note: Resembles frost damage in young crops, although in this case there is likely to be bronzing of leaves as well as buckling.

226

227

228

Magnesium deficiency

Interveinal chlorosis (parallel yellowish-white stripes between green veins) of older leaves, followed by red and purple tints on edges and tips of these leaves, which may die back.

228 Affected maize leaves: tip section at bottom shows reddening and die-back.

Notes:
1. Streak virus can be confused with magnesium deficiency but is confined to Africa and south east Asia (Bock, 1974).
2. See notes on iron deficiency.
3. Triazine-type herbicides can cause a similar interveinal chlorosis.

Iron deficiency

Interveinal chlorosis (striping) on entire length of new leaves.

229 Right, deficient leaf; left, leaf showing necrotic blotches resulting from an excess of iron.

Note: Resembles magnesium deficiency in its early stages but becomes progressively worse on new leaves, which may be completely bleached. Occurs very readily in artificial growth media.

Manganese deficiency

Collapse of mesophyll tissue, producing white streaks between green veins. In severe cases the centres of these streaks may turn brown and drop out. Rare. (Piper, 1940a, Muller, 1973).

(Not illustrated)

229

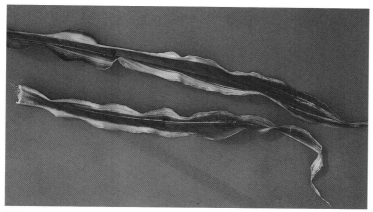

230

Zinc deficiency

Broad bands of whitish-coloured tissue appear in the lower half of emerging leaves in the young crop, especially in cool, damp weather. In mild cases the bands may appear as several interveinal stripes, but these do not extend the full length of the leaf as with iron and magnesium deficiency. This symptom is known as 'white bud' in USA. Anthers may be missing from tassels and emergence of silks is delayed, which may reduce yield due to poor pollination. Reddish-brown discoloration at nodes in stem. Shortened internodes.

230 Leaf symptoms in maize.

Note: See Lingle and Holmberg (1957) for further details.

Boron deficiency

Distortion of spikelets; absence of anthers; barren or partly barren ears, especially at tip which is pointed. Irregular white spots on youngest leaves which coalesce to form raised, waxy stripes up to 5 cm long. Leaves thick; midribs brittle; internodes short.

231 Three deficient cobs showing bare tips and irregular and incomplete arrangement of grains.

Note: See Shorrocks and Blaza (1973) for further details.

231

Watercress

232

233

234

Healthy plant

232 Healthy plants — note green petioles.

Nitrogen deficiency

Uniform chlorosis of older leaves. Pink petioles and veins.
233 Left to right: leaves of decreasing age.

Note: Rare in commercial beds but once recorded in UK (Hamence and Taylor, 1947) where nitrate concentration in water was below detection limit.

Phosphorus deficiency

Stems, petioles, veins purple. Leaves dark green, bronzed, especially on underside and small. Shortened internodes.
234 Left, underside of old leaf; centre, upper surface of old leaf; right, upper surface of young leaf.

Note: Exacerbated by poor root system, as with 'crook-root' (Tomlinson, 1960) or poor attachment to substrate.

235

236

237

Potassium deficiency

Marginal chlorosis and scorch of older leaves.

235 Left and centre old leaves; right younger leaf. Rare.

Iron deficiency

Interveinal chlorosis of young leaves.

236 Left to right: leaves of increasing age.

Note: Iron deficiency is readily induced in watercress by excessive application of phosphatic fertilizer (Robinson and Cumbus, 1977) or whenever there is poor root attachment to the substrate, thus removing the prime source of iron (See also manganese deficiency notes).

Manganese deficiency

Interveinal chlorosis of young leaves, almost indistinguishable from iron deficiency.

237 Right, young leaf: centre and left, mature leaves.

Note: Cumbus and Robinson (1977) have shown that chlorosis in well-rooted watercress is associated with leaf Mn concentrations below 20 $\mu g\ g^{-1}$ and Zn concentrations over 100 $\mu g\ g^{-1}$. The latter may result from excessive use of zinc to control 'crook-root'.

APPENDIX 1
Inducing and illustrating plant mineral disorders

Almost all of the photographs in the Colour Atlas which have the words 'sand culture' in the caption, and all those resulting from 'peat culture' were obtained from pot experiments done at NVRS during the spring of 1978 and 1979. These are referred to throughout the book as 'the NVRS experiments' and some details of these and the photographic methods used are given below. First, however, it may be of value to anyone contemplating a similar exercise to discuss briefly the various approaches which are possible.

To begin with, there is the question of photography versus hand painting of subjects. In the past, the latter has usually been preferred for botanical illustration. It has the advantage that irrelevant features can be suppressed and important ones emphasised: also there are no problems with narrow planes of focus or avoiding distracting backgrounds as with photography. (A good example of the technique is the book by Gram et al., 1958). On the other hand, an artist is unlikely to be available to work from the live subject, which means that a fairly good photograph will be required anyway. This being so, the photographer might as well try to obtain publishable results straight away.

The subject plants would ideally come from affected field-grown crops for which the diagnosis is beyond any doubt. In a country where most growers are taking precautions to avoid mineral imbalances this approach is obviously slow to produce results, although an advisory officer who is also a photographer should have excellent opportunities to acquire a good collection of pictures in this way. (However, he would be ill-advised to rely entirely on photographing plants in the field because subject movement caused by wind, and poor light, can combine to prevent good results from being obtained, unless flash is used.) Field experiments are another possibility but even if a field has been deliberately 'run down' to induce deficiencies, it is unlikely that more than two or three major-element deficiencies will occur. For others one is likely to have to select sites at some considerable distance away, which greatly complicates the problems of observation and photography. Pot-grown plants have the advantage that they can be taken into a studio to be photographed, and then returned to the glasshouse to allow symptoms to develop further. However, if pure sand is used to provide a nutrient-free supporting medium for the plants, as was done by Wallace (1961) and by ourselves for the six major nutrients, it is quite difficult to arrange for the deficiency to develop gradually and at about the same growth stage as typically occurs in the field. Workers at GCRI have found peat to be a useful growing medium for the purpose, and we succeeded in inducing boron deficiency symptoms very readily this way (see below). It would appear that by using a growing medium like this which has a great ability to adsorb and precipitate micro-elements, it may not always be necessary to take strict precautions to exclude all traces of these elements from the nutrient solution, as was done by Wallace and his co-workers. After all, this is more akin to the way in which deficiencies occur in the field situation. Another possibility is to collect soil from known areas of deficiency, and carry out pot experiments with them at a central site. Although this method would appear to have much to commend it, we were unable to induce copper or manganese deficiencies in this way on a peat soil from Methwold Fen, normally regarded as deficient in these elements. The photographs of micro-element deficiencies in the Colour Atlas for which the words 'flowing solution culture' appear in the caption were obtained by Dr Hewitt and Mr Watson at Long Ashton Research Station, who used a system of plastic guttering to conduct solutions, each lacking one nutrient, from a header tank to a lower reservoir. They found that plants grown in such channels show symptoms identical to those seen in sand or deep water culture, but they cannot be removed to a studio for photography because the roots become inextricably entwined with those of neighbouring plants. Whatever system of plant support and growing medium is used, there is no doubt that a typical greenhouse environment produces a rather different type of plant, generally less pigmented, than the outdoor environment. Therefore it seems preferable to conduct such experiments out of doors, with precautions to avoid damage by wind, birds etc. Probably the ideal environment is the mobile type of greenhouse which is popular in Germany. This can be used to provide protection from wind, rain and frost, but can be rolled away from the plants at other times. If a conventional glasshouse is used, it is important that night temperatures should be low (but with frost protection) and that shading be avoided as far as possible.

Turning to the NVRS experiments, both of these were of 'minus one' design, in which a complete nutrient solution, applied at almost every watering, is compared with a number of others in each of which a single nutrient element is absent. Both experiments were carried out in a heated glasshouse, starting in midwinter and ending in early summer. Supplementary illumination in the form of 400W HLRG lamps was provided from 2300 h until 0900 h initially, (0001 h until 0930 h in 1979) but this was dispensed with in early March (early April in 1979). Temperatures varied from 10°C at night to 30°C on sunny days. The species used, and their cultivars, are given in Table 2. 12.5 cm plastic pots, standing on slatted wooden benches, were used for both experiments. The 'complete, full strength' nutrient solution for both experiments, consisted of the following salts at the concentrations stated

Broad bean	Conqueror
Broccoli	Purple sprouting
Brussels sprout	Ulysses
Cabbage	Summer Monarch
Carrot	Chantenay Red Cored
Cauliflower	Finney's 110 (All the year round in 1979)
Celery	New dwarf white
Curly kale	Dwarf green curled (omitted in 1979)
French bean	The Prince (Tendergreen in 1979)
Leek	Catalina
Lettuce	Cobham green
Marrow	Zucchini (Chefini in 1979)
Onion	Rijnsburger
Parsnip	Improved marrow
Pea	Dark skinned perfection
Radish	Cherry Belle
Red beet	Boltardy
Spinach	Clean leaf (Sigmaleaf in 1979)
Swede	Western Perfection
Sweet corn	Earliking
Turnip	Snowball

Table 2 Vegetable cultivars used in NVRS experiments (unless otherwise stated, the same cultivars were used for both experiments)

(in mM): KNO_3 4.0; $Ca(NO_3)_2$ 3.0; $MgSO_4$ 0.75; NaH_2PO_4 4.0; $NaNO_3$ 1.5; $MnSO_4$ 0.01; $CuSO_4$ and $ZnSO_4$ 0.001; H_3BO_3 0.03; $(NH_4)_6 Mo_7O_{24}$ 0.00003; Fe EDTA (Na salt) 0.078. The salts used were hydrated, Analar grade, and were dissolved in deionised rainwater.

The 1978 (major nutrient) experiment made use of acid-washed sand supplied by Long Ashton Research Station for the minus Ca and minus S treatments, plus a set of complete nutrient controls, and water-washed silver sand for the remainder (minus N, P, K, Mg and a further set of controls). The minus N treatment was obtained by replacing the KNO_3 and $Ca(NO_3)_2$ of the complete solution by KCl and $CaCl_2$ respectively, and omitting $NaNO_3$. For minus P, NaH_2PO_4 was omitted: for minus S, $MgSO_4$ was replaced by $MgCl_2$ (the micro-element sulphates were not replaced): for minus K, KNO_3 was replaced by $Ca(NO_3)_2$: for minus Ca, $Ca(NO_3)_2$ was replaced by KNO_3, and for minus Mg, $MgSO_4$ was replaced by K_2SO_4. In all cases, replacement was on a mol for mol basis. The solutions were all used at half strength until late March. Deionised water was used at weekends except in the later stages of the experiment. There were two series of pots: series one received the treatment solutions from the outset, except for small-seeded crops which received a one eighth strength complete nutrient solution for the first watering, in case deficiency should appear too soon, whereas series two received the complete solution until late March, after which treatment solutions were phased in on each species as it achieved a fairly advanced stage of growth. Seeds, numbering one to six per pot according to species, were sown on 1st February. After emergence, plants were gradually thinned out, leaving one per pot for large plants (peas, beans, sweet corn) to two for the smallest. On series one, expected deficiency symptoms began to appear about 16 days after

sowing on the minus Ca treatment, 20 days on minus S, 22 days on minus N, 27 days on minus Mg, 29 days on minus K and 34 days on minus P. Reduction of growth, recorded as ground cover measurements on 1st March, was greatest (about 80%) on the minus N treatment (except for peas and broad beans), almost as great on minus S treatments, but fairly small on the other treatments. Indeed, some brassica crops appeared larger on minus P treatments than on the complete nutrient solution.

On series two, which received treatment solutions from 23rd March, symptoms were first noted on 4th April. By mid-May all treatments had resulted in symptoms on most species, except that on the minus P treatment growth was often better than on the controls. The reason for this was not clear and no detectable P was found in either the nutrient solution or the leachate from the sand.

As symptoms on both series developed, the plants were taken to the NVRS studio where they were photographed using a 6×6 cm format single lens reflex camera, two Bowen 'Monolite' tungsten lamps, and Agfachrome 50L transparency film, against a black velvet background. Three exposures were taken of each subject, one stop apart, and a Kodak colour 'wedge' was held in the plane of focus, on the edge of the picture area, to provide a check on the correctness of colours and a length scale on the transparency. In many instances the leaves which were photographed were immediately removed and dried for analysis: corresponding leaves from healthy control plants were collected and dried within 2–3 days for comparison. Analytical methods used were those of MAFF (1973), except that atomic absorption spectrophotometry was used to determine calcium and magnesium. ('Total' nitrogen figures do *not* include nitrate-N.)

For the 1979 micronutrient experiment, Irish sphagnum peat was used, limed to a pH of 6.8. Pot size, growing conditions etc were as above. The treatments consisted of omissions of boron, manganese and copper plus a complete nutrient control, and were achieved by excluding the appropriate salts from the nutrient solution already specified. The peat used for the minus Mn and minus Cu treatments had already been used for a similar experiment and by superimposing the same treatments for a second time on the already depleted peat it was hoped to induce these deficiencies, even though they had not been observed at the first attempt.

There were four replications, and the treatment solutions were all applied, but at half strength, to all replicates from the day of sowing, 17th January. However, because symptoms of boron deficiency were already apparent on some seedlings on 12th February, two replicates of the minus B treatment were put back onto complete (half strength) solution until 20th March, on which day a 'crossover' was done, with plus B replicates becoming minus B and vice-versa. From mid-March, all solutions were applied at full strength.

As mentioned, boron deficiency symptoms appeared on the minus B treatment very early, especially on cauliflower, spinach and turnip, although the turnip then recovered. Apart from stunting of red beet and slight cupping of Brussels sprouts, no further symptoms appeared until 6th March, when radish roots were found to have split. By early April, all species which had received minus B solutions in early life (except cabbage and sweet corn) were showing marked B deficiency symptoms, and growth was reduced to between a quarter and three quarters of that of control plants. Cauliflowers were still barely beyond the cotyledon stage (see Plate 31, p. 28) whereas cabbages were as large as controls, illustrating the great variation in susceptibility within the Brassica group. (Later, these cabbages rosetted.) Throughout the experiment sweet corn showed a chlorosis and bleaching thought to be iron deficiency on all treatments, but no other symptoms. Copper deficiency symptoms failed to appear on any crop, and manganese deficiency only affected purple sprouting broccoli. By mid-May several crops on all treatments were showing signs of calcium deficiency and the experiment was terminated.

During this experiment, photographs were taken as before, but no plant analysis was done.

Early in 1982 it was found that crops grown in Somerset (sedge) peat limed to pH 7.0 exhibited clear symptoms of manganese deficiency. These were confirmed by leaf analysis and by spraying affected plants with 0.04 M $MnSO_4$, which resulted in a complete cure. Plates 26, 73, 159 and 175 and accompanying data were obtained in this way.

Some observations on trace element deficiencies in vegetables

These observations were made at Lullymore, Co Kildare, Eire and contributed by F. S. MacNaeidhe.

The research station soil at Lullymore consists of 'cutover peatland' typical of much of the Irish midlands. In the undisturbed raised bog, an upper layer of relatively undecomposed Sphagnum peat overlies a more decomposed layer of Sphagnum/Eriophorum mixture, with at least a metre of fen peat below that. The cutover land has had most of the top two layers removed for fuel.

The primary nutrient on this land when newly brought into cultivation is copper, without which no vegetables will grow. However, a soil dressing of 44 kg ha^{-1} copper will suffice for most crops for up to 10 years. Soil levels should be 5–10 μg g^{-1} (EDTA extract). When the level in the foliage drops below 3 μg g^{-1}, deficiency symptoms will almost certainly show up. The copper response is not sensitive to pH.

The boron requirement, on the other hand, is very pH dependent. B must be applied in a foliar spray to carrots and celery if soil pH exceeds 6.0. Two or three sprays each containing 4.4 kg ha^{-1} 'Solubor' (20.5% B) are used. These crops will grow vigorously up to pH 7.0 so long as rain falls, but drought tends to induce B deficiency in carrots at a pH above 6.5. This takes the form of (a) a prostrate growth habit, (b) absence of leaf tissue—'fern leaf' symptoms and (c) root splitting. At a similar pH celery will remain unaffected even during severe drought but when the weather breaks severe B deficiency symptoms can occur, resulting in complete crop failure at pH values of over 6.5. The typical symptoms such as 'cat's claw' and 'brown streak' do not always appear, but the growing point breaks down into a brown mass. The symptoms look very much like 'black heart' except that the colour of the growing point is brown. The condition begins to occur within 2–3 days of the end of a severe drought.

Iron, manganese and zinc deficiency normally never occur unless pH is abnormally high. Iron deficiency has not been seen in vegetable crops, but French bean and onion suffer from manganese deficiency above pH 6.5, carrots above 7.0 and potatoes above 7.5. In carrots the foliage remains upright but stunted and chlorotic: french beans show interveinal chlorosis, and onions simple (uniform) chlorosis. Drought tends to increase the severity of manganese deficiency.

Zinc is also pH sensitive but in contrast to manganese is much more severe when the weather is wet, overcast and abnormally cold. Under these conditions the zinc should be applied as a routine foliar spray in french beans, onions and sweet corn when the pH is greater than 6.0. In sweet corn cold symptoms and zinc deficiency symptoms seem to be synonymous, appearing as yellow streaks on the leaves, and the response to zinc spray is very poor or absent depending on the temperature and light intensity. Zinc deficiency is less tempture/light dependent in onions and appears as a yellowing and twisting of the foliage. Occasional plants show yellow and green streaks.

Molybdenum deficiency has not been recorded in crops on peat soil here at a pH of 5.4 or greater. Cauliflower seed should be dressed as a routine if soil is below this pH; below pH 5.3 few cauliflowers will emerge unless the seed is dressed with molybdenum. At a pH of 5.2 cabbage seedlings became chlorotic and died out on peat soil but have grown successfully at this pH when the seed was dressed. Seed dressing with molybdenum has improved the percentage of emergence in onions up to pH 5.4 and poor emergence and death of seedlings occurs commonly at a pH of 5.0 or less if seed is not dressed. Individual onion plants will survive on peat at a pH of 4.2 but these tend to be of very poor quality. A temperature/light/moisture factor appears to be involved in the availability of molybdenum also. At Lullymore we have failed to reproduce molybdenum deficiency symptoms in some peat types in the glasshouse.

It is very difficult to be quite definite about pH values in the soil and trace element deficiencies, but a value between 5.3 and 6.0 should be regarded as ideal. I have, however, seen quite excellent crops of carrots, cauliflowers, celery, french beans and onions grown at a pH of 7.0 or more without symptoms of trace element deficiencies. On the other hand, zinc deficiency has occurred in onions in one case at a pH of 5.8. I have also seen a healthy onion crop growing at a pH of 4.8 without the benefit of a seed dressing of molybdenum. However, what I have written is a summary of the general guidelines which we follow ourselves and which has given satisfactory results with growers on peat soil.

Acknowledgements of photographs contributed

The authors express their grateful thanks to the following individuals and organisations who have generously provided the photographs for the plate numbers shown:

Agricultural and Development Advisory Service, UK: Plates 81, 109, 117, 172, 204 and p. 12 (Crown copyright.)

W. Bussler, Technical University, Berlin, FRG: Plates 9, 13, 44, 88, 181, 231.

Centro Internacional de Agriculture Tropical (CIAT), Cali, Columbia: Plate 183.

I. P. Cumbus, Oxford Polytechnic: Plates 232, 233, 234, 235, 236, 237.

E. J. Hewitt and E. Watson, Long Ashton Research Station, UK: Plates 25, 28, 29, 56, 58, 59, 74, 128, 157, 160, 173, 176, 177, 215, 217, 218, 220.

Kali und Sulz A. G., Kassel, FRG: Plate 226.

H. Kühn, Justus-Liebig University, Giessen, FRG: Plate 153.

R. E. Lucas, University of Florida, USA: Plate 111.

F. S. MacNaeidhe, The Agricultural Institute, Eire: Plate 75.

D. N. Maynard, University of Florida, USA: Plates 18, 42, 170, 225, 229.

E. R. Page, National Vegetable Research Station, UK: Plate 19.

Phosyn Chemicals Ltd., Pocklington, York: Plate 230.

R. Piot, Pedology Service, Belgium: Plates 192, 216.

D. T. Pope and the Fertiliser Institute, Washington, USA: Plate 91.

Processors and Growers Research Organisation, UK: Plate 156, 158.

C. A. Shadbolt, Kansas, USA: Plate 222.

K. W. Smilde, Roorda van Eysinga and PUDOC (Netherlands): Plates 110, 113, 116.

D. A. Stone, National Vegetable Research Station, UK: Plate 114.

All other plates are from photographs taken at the National Vegetable Research Station, Wellesbourne, Warwick, UK, (which retains the copyright) or from Wallace (1961).

Glossary

Abaxial: the side or face away from the stem; the lower surface of a leaf

Absciss: fall off

ADAS: Agricultural Development and Advisory Service of the Ministry of Agriculture, Fisheries & Food, UK

Adaxial: the side or face next to the stem; the upper surface of a leaf

Adsorption capacity: capacity of soil to attract and hold ions on its surface

Anion: negative ion eg Cl^-, SO_4^{--}, PO_4^{---}

Apical dominance: the suppression of growth of side shoots under hormonal control from the apex (tip of main shoot)

Axillary: side (shoots). Shoots growing in angle of main stem and leaf

B: Boron

Blind growing point: growing point absent—not necrotic

Bractiness: production of scale-like leaves on the surface of cauliflower curd

Ca: Calcium

Cation: positive ion eg Na^+, K^+, NH_4^+, Mg^{++}, Ca^{++}

Chlorosis: synonymous with 'yellowing': uniform over whole leaf unless said to be interveinal

Chlorotic marbling: a blotchy yellowing of leaves in which the yellow merges softly into the green

CIAT: International Centre for Tropical Agriculture, PO Box 67–13, Cali, Colombia

Cl: Chlorine

Concave: bulging downwards in centre, when seen from above

Convex: bulging upwards in centre, when seen from above

Cotyledons: seed leaves

Critical concentration: see p. 15

Cu: Copper

Cytokinins: plant growth substances involved in cell division

Denitrification: microbial conversion of mineral nitrogen to nitrogen gas, under anaerobic conditions

Distal: (part of leaf) furthest from stem

Epidermis: plant skin

Fasciation: widening and flattening of stem

Fe: Iron

Fortnight: two weeks (USA)

GCRI: Glasshouse Crops Research Institute, Rustington, Littlehampton, West Sussex, UK

Hearting: heading up (USA)

Hypocotyl: main stem of embryo below the cotyledons which develop into root

Inter: between

Internode: part of stem between two joints (nodes)

Interveinal: between the veins of the leaf

Intra: within

Ion: electrically charged particle

K: Potassium

Lamina: blade of leaf

LARS: Long Ashton Research Station, University of Bristol, Bristol, U. K.

Lattice: the term used in soil chemistry to denote the micro-structure of clay minerals, which consists of layers of alumina and silica

Leaching: the washing out of mineral salts and organic matter from one layer of soil into a lower layer by percolating rain water

MAFF: Ministry of Agriculture, Fisheries and Food, UK

Major or Macronutrients: those required in relatively large amounts for plant growth, ie N, P, S, K, Ca, Mg

Marginal: around leaf edges

Mature leaves: fully expanded but not senescent

Mesophyll collapse: tissue collapse—the affected area takes on a dull, limp appearance as in Illustration 17

Mg: Magnesium

Minor or Micronutrients: those required in relatively small amounts for growth, ie Fe, Mn, Zn, Cu, B, Mo

Midrib: main vein of leaf

Mineralisation: microbial conversion of organic materials to inorganic (mineral) forms, eg protein to nitrate

Mn: Manganese

Mo: Molybdenum

Monocotyledons: narrow leaved plants eg grasses, cereals

N: Nitrogen (Note that 'Total N' figures quoted do not include nitrate)

Necrosis: tissue death

Nematodes: eelworms

Nitrate: Nitrate figures all refer to NO_3, (not NO_3-N) in fresh petiole sap, extracted as described in chapter two

NO_3: Nitrate

Node: joint where leaf is borne

Nodulation: formation of root nodules containing nitrogen-fixing bacteria

NVRS: National Vegetable Research Station, Wellesbourne, Warwick, UK

Pathogen: disease producing organism (fungus or virus)

P: Phosphorus

Petiole: leaf stalk

PGRO: Pea Growing Research Organisation Ltd, The Research Station, Great North Road, Thornhaugh, Peterborough, UK

pH: index of acidity (usually soil)—actually negative logarithm of hydrogen ion activity. All references to pH apply to a 1:2 soil: water (v/v) suspension

Physiological age: plant age expressed as degree of development of leaves, flowers etc, rather than in time units

Physiologically active: not showing signs of senescence

Phytotoxic: poisonous to plants

Proximal: (part of leaf) nearest the stem

Pulvinus: swollen base of leaf stalk

S: Sulphur

Scorch: death and drying out of tissue

Subclinical deficiency: range of deficiency in which symptoms are not seen but growth is reduced

Suberisation: formation of cork

Tipburn: death and browning of a narrow marginal band of leaf, often preceded by the tissue concerned becoming limp and exhibiting brown spots and/or discoloured veins. The term is usually applied to symptoms on relatively young leaves, enclosed or partly enclosed by outer wrapper leaves

Upper critical concentration: See p. 15

Vascular system: network of conducting vessels from roots to leaves, including veins, midribs etc

Xylem: the main water-conducting plant tissue

Zn: Zinc

References

Ashkar, S. A. and Ries, S. K. (1971). Lettuce tipburn as related to nutrient imbalance and nitrogen composition. *J Am Soc hort Sci* **96,** 448–452.

Atkin, J. D. (1958). Relative susceptibility of snap bean varieties to mechanical injury of seed. *Proc Am Soc hort Sci* **72,** 370–373.

Beaumont, A. B. and Snell, M. E. (1935). The effect of magnesium deficiency on crop plants. *J agric Res* **50,** 553–562.

Berger, K. C. (1962). Micronutrient deficiencies in the United States. *J agric Fd Chem* **10,** 178–181.

Blank, F. (1947). The anthocyanin pigments of plants. *Bot Rev* **13,** 241–317.

Bock, K. R. (1974). Maize streak virus. *CMI/AAB Descr Pl Viruses* **133,** 4 pp.

Brenchley, W. E. and Warington, K. (1927). The role of boron in the growth of plants. *Ann Bot* **41,** 167–188.

Bussler, W. (1958). Manganvergiftung bei höheren Pflanzen. *Z Pflernähr Düng* **81,** 256–265.

Chapman, H. D. (1966). Diagnostic criteria for plants and soils. *Berkeley Univ Calif Div Agric Sci* 793 pp.

Chupp, C. and Sherf, A. F. (1960). Vegetable diseases and their control. *New York, Ronald P* 693 pp.

Cumbus, I. P. and Robinson, L. W. (1977). Trace element imbalance in watercress (*Rorippa nasturtium-aquaticum* (L) Hayek). *Hort Res* **16,** 57–60.

Duffus, J. E. (1972) Beet western yellows virus. *CMI/AAB Descr Pl Viruses* **89,** 4 pp.

English, J. E. and Maynard, D. N. (1978). A key to nutrient disorders of vegetable plants. *Hort Science* **13,** 28–29.

Geraldson, C. M. (1954). The control of blackheart of celery. *Proc Am Soc hort Sci* **63,** 353–358.

Geraldson, C. M., Klacan, G. R. and Lorenz, O. A. (1973). Plant analysis as an aid in fertilising vegetable crops. *In* Walsh, L. M. and Beaton, J. D. eds. Soil testing and plant analysis. Madison, Wisconsin, *Soil Sci Soc Am* 491 pp.

Gibbs, A. J. and Harrison, B. D. (1970). Cucumber mosaic virus. *CMI/AAB Descr Pl Viruses* **no 1** 4 pp.

Gram, E., Bovien, P. and Stapel, C. (1958). Recognition of diseases and pests of farm crops. Cambridge: *Danish Agricultural Information and Advisory Aids Service* 128 pp.

Greenwood, D. J., Barnes, A., Liu, K., Hunt, J., Cleaver, T. J. and Loquens, S. M. H. (1980). Relationships between the critical concentrations of N, P and K in 17 different vegetable crops and duration of growth. *J Sci Fd Agric* **31,** 1343–53.

Gupta, U. C. (1979). Boron nutrition of crops. *Adv Agron* **31,** 273–303.

Gupta, U. C. and Cutcliffe, J. A. (1973). Boron nutrition of broccoli, brussels sprouts, and cauliflower grown on Prince Edward Island soils. *Can J Soil Sci* **53,** 275–279.

Hamence, J. H. and Taylor, G. (1947). Nitrate deficiency in water-cress. *Agriculture (Lond)* **54,** 358–361.

Hewitt, E. J. (1944–1951). Rep Long Ashton Res Stn 1943–50.

Hewitt, E. J. and Smith, J. A. (1975). Plant mineral nutrition. *London, English Univ P* 298 pp.

Hohlt, H. E. and Maynard, D. N. (1966). Magnesium nutrition of spinach. *Proc Am Soc hort Sci* **89,** 478–482.

International Lead Zinc Research Organisation, (nd). *Zinc in crop nutrition,* 64 pp.

Johnson, C. M., Pearson, G. A. and Stout, P. R. (1952). Molybdenum nutrition of crop plants. II. Plant and soil factors concerned with molybdenum deficiencies in crop plants. *Pl Soil* **4,** 178–196.

Johnson, C. M., Stout, P. R., Broyer, T. C. and Carlton, A. B. (1957). Comparative chlorine requirements of different plant species. *Pl Soil* **8,** 337–353.

Kawaski, T. and Moritsugu, M. (1979). A characteristic symptom of calcium deficiency in maize and sorghum. *Commun Soil Sci Pl Anal* **10,** 41–56.

Krantz, B. A. and Melsted, S. W. (1964). Nutrient deficiencies in corn, sorghums, and small grains. *In H. B. Sprague,* ed. *Hunger signs in crops.* **3rd ed 25–57.** New York, David McKay Company.

Lingle, J. C. and Holmberg, D. M. (1957). The response of sweet corn to foliar and soil zinc applications on a zinc deficient soil. *Proc Am Soc hort Sci* **70,** 308–315.

Maynard, D. N., Gersten, B., Vlach, E. F. and Vernell, H. F. (1961). The effects of nutrient concentration and calcium levels on the occurrence of carrot cavity spot. *Proc Am Soc hort Sci* **78,** 339–342.

Millikan, C. R. and Hanger, B. C. (1966). Calcium nutrition in relation to the occurrence of internal browning in brussels sprouts. *Aust J agric Res* **17,** 863–874.

Ministry of Agriculture, Fisheries and Food (1973). The analysis of agricultural materials. *Tech Bull Minist Agric, Fish Fd* **no 27.**

Ministry of Agriculture, Fisheries and Food (1976). Trace element deficiencies in crops. *ADAS adv pap* **no 17,** 34 pp.

Ministry of Agriculture, Fisheries and Food (1979). Fertiliser recommendations. *Ministry of Agriculture, Fisheries and Food,* **GF1,** 92 pp.

Mulder, E. G. (1954). Molybdenum in relation to growth of higher plants and micro-organisms. *Pl Soil* **5,** 368–415.

Muller, G. J. (1973). A compendium of corn diseases. *St Paul, Minnesota, Am Phytopath Soc.*

Needham, P. (1971). Cavity spot of carrots. *MAFF/ADAS plant physiology committee* **SS/C/304.**

Needham, P. (1976). Plant analysis data. *MAFF.* **SS/Phys/67.**

Piper, C. S. (1940a). The symptoms and diagnosis of minor-element deficiencies in agricultural and horticultural crops. I. Diagnostic methods. Boron. Manganese. *Emp J exp Agric* **8,** 85–96.

Piper, C. S. (1940b). The symptoms and diagnosis of minor-element deficiencies in agricultural and horticultural crops. II. Copper, Zinc, Molybdenum. *Emp J Exp Agric* **8,** 199–206.

Pizer, N. H., Caldwell, T. H., Burgess, G. R. and Jones J. L, O. (1966). Investigations into copper deficiency in crops in East Anglia. *J agric Sci, Camb* **66,** 303–314.

Plant, W. (1951). The control of 'whiptail' in broccoli and cauliflower. *J hort Sci* **26,** 109–117.

Plant, W. (1956). The effects of molybdenum deficiency and mineral toxicities on crops in acid soils. *J hort Sci* **31,** 163–176.

Purvis, E. R. and Ruprecht, R. W. (1937). Cracked stem of celery caused by a boron deficiency in the soil. *Fla Agr Exp Sta Bul* **307,** 16 pp.

Robinson, L. W. and Cumbus, I. P. (1977). Determination of critical levels of nutrients in watercress (*Rorippa nasturtium-aquaticum* (L) Hayek) grown in different solution concentrations of N, P and K. *J hort Sci* **52,** 383–390.

Scaife, M. A. and Bray, B. G. (1977). Quick sap tests for improved control of crop nutrient status. *ADAS q Rev* **no 27,** 137–145.

Seeliger, M. T. and Moss, D. E. (1976). Correction of iron deficiency in peas by foliar sprays. *Aust. J. exp. Agric. Anim. Husb.,* **16,** 758–760.

Shadbolt, C. A. (1959). Nitrogen fertilisation and spacing as related to shrivel in sweet corn. *Proc Am Soc hort Sci* **74,** 446–453.

Shannon, S., Natti, J. J. and Atkin, J. D. (1967). Relation of calcium nutrition to hypocotyl necrosis of snap bean (*Phaseolus vulgaris* L). *Proc Am Soc hort Sci* **90,** 180–190.

Shepherd, R. J. (1970). Cauliflower mosaic virus. *CMI/AAB Descr Pl Viruses* **no 24,** 4 pp.

Shorrocks, V. M. (1974). Boron deficiency—its prevention and cure. *London: Borax consolidated Ltd.,* 56 pp.

Shorrocks, V. M. and Blaza, A. J. (1973). The boron nutrition of maize. *Wild Crops* **25,** 25–27.

Stenuit, D. and Piot, R. (1960). Symptomes de carence et de toxicité en manganese chez les plantes agricoles et horticoles. *Serv pédol Belg.* 47 pp.

Stuart, N. W. and Griffin, D. M. (1944). Some nutrient deficiency effects in the onion. *Herbertia, La Jolla* **11,** 329–337.

Thomas, M. D. (1951). Gas damage to plants. *A Rev Pl Physiol* **2,** 293–322.

Tomlinson, J. A. (1960). Crook-root disease of watercress–a review of research. *NAAS q Rev* **4,** 13–19.

Tomlinson, J. A. and Ward, C. M. (1978). The reactions of swede (*Brassica napus*) to infection by turnip mosaic virus. *Ann appl Biol* **89,** 61–69.

Tomlinson, J. A. and Webb, M. J. W. (1978). Ultrastructural changes in chloroplasts of lettuce infected with beet western yellows virus. *Physiol Pl Path* **12,** 13–18.

Tompkins, D. R., Baker, A. S., Gabrielson, R. L. and Woodbridge, C. G. (1965). Sulfur deficiency of broccoli. *Pl Dis Reptr* **49,** 891–894.

Van Eysinga, N. L. R. and Smilde, K. W. (1971). Nutritional disorders in glasshouse lettuce. *Wageningen, Centre for Agricultural Publishing and Documentation.* 56 pp.

Vitosh, M. L., Warncke, D. D. and Lucas, R. E. (1973). Secondary and micronutrients for vegetables and field crops. *Ext Bull Mich St Univ agric Exp Stn, E–486,* 19 pp.

Walker, J. C. and Edgington, L. V. (1957). Studies of internal tipburn of cabbage. *Phytopathology* **47,** 537.

Wallace, T. (1961). The diagnosis of mineral deficiencies in plants by visual symptoms. *London, HMSO.*

Watson, M., Serjeant, E. P. and Lennon, E. A. (1964). Carrot motley dwarf and parsnip mottle viruses. *Ann appl Biol* **54,** 153–166.

Yamaguchi, M. and Minges, P. A. (1956). Brown checking of celery, a symptom of boron deficiency. I. Field observations, varietal susceptibility, and chemical analysis. *Proc Am Soc hort Sci* **68,** 318–328.

Yamaguchi, M., Takatori, F. H. and Lorenz, O. A. (1960). Magnesium deficiency in celery. *Proc Am Soc hort Sci* **75,** 456–462.

Zink, F. W. (1966). The response of head lettuce to soil application of zinc. *Proc Am Soc hort Sci* **89,** 406–414.

Index

Printed in the UK for HMSO
Dd 718755 C50 9/83 29027